ORGANIC FARMING
An International History

In memory of Ben Stinner

His insightful intelligence, quick wit and engaging geniality delighted and inspired all who knew him

Organic Farming

An International History

Edited by

William Lockeretz
Friedman School of Nutrition Science and Policy, Tufts University, Boston, Massachusetts, USA.

CABI is a trading name of CAB International

CABI Head Office
Nosworthy Way
Wallingford
Oxfordshire OX10 8DE
UK

CABI North American Office
38 Chauncey St
Suite 1002
Boston, MA 02111
USA

Tel: +44 (0)1491 832111
Fax: +44 (0)1491 833508
Email: cabi@cabi.org
Web site: www.cabi.org

Tel: +1 617 395 4051
Fax: +1 617 354 6875
Email: cabi-nao@cabi.org

©CAB International 2007. All rights reserved. No part of this publication may be reproduced in any form or by any means, electronically, mechanically, by photocopying, recording or otherwise, without the prior permission of the copyright owners.

A catalogue record for this book is available from the British Library, London, UK

Library of Congress Cataloging-in-Publication data
Organic farming: an international history / William Lockeretz, editor.
   p. cm.
   ISBN 978-0-85199-833-6 (alk. paper) – ISBN 978-1-184593-289-3 (ebook)
 1. Organic farming—History. 2. Organic farming—Societies, etc. I. Lockeretz, William.
II. Title.

S605.5.O667 2007
631.5'84—dc22

2007012464

ISBN-13: 978-0-85199-733-6 (HB)

ISBN-13: 978-1-84593-876-5 (PB)

First published (HB) 2007
First paperback edition 2011
Reprinted 2014

Printed and bound in the UK by CPI Group (UK) Ltd, Croydon, CR0 4YY

The paper used for the text pages of this book is FSC certified. The FSC (Forest Stewardship Council) is an international network to promote responsible management of the world's forests.

Contents

Contributors	vii
Foreword Nadia El-Hage Scialabba	ix

I: Origins and Principles

1.	**What Explains the Rise of Organic Farming?** W. Lockeretz	1
2.	**The Origins of Organic Farming** G. Vogt	9
3.	**Organic Values** M. Sligh and T. Cierpka	30
4.	**The Science of Organic Farming** D.H. Stinner	40
5.	**The Evolution of Organic Practice** U. Niggli	73

II: Policies and Markets

6.	**The Development of Governmental Support for Organic Farming in Europe** S. Padel and N. Lampkin	93

v

7.	**The Organic Market** *J. Aschemann, U. Hamm, S. Naspetti and R. Zanoli*	123
8.	**Development of Standards for Organic Farming** *O. Schmid*	152

III: Organizations and Institutions

9.	**IFOAM and the History of the International Organic Movement** *B. Geier*	175
10.	**The Soil Association** *P. Conford and P. Holden*	187
11.	**Ecological Farmers Association and the Success of Swedish Organic Agriculture** *I. Källander*	201
12.	**MAPO and the Argentinian Organic Movement** *D. Foguelman*	217
13.	**NASAA and Organic Agriculture in Australia** *E. Wynen and S. Fritz*	225
14.	**FiBL and Organic Research in Switzerland** *U. Niggli*	242
15.	**The Organic Trade Association** *K. DiMatteo and G. Gershuny*	253

IV: Challenges

16.	**A Look Towards the Future** *B. Geier, I. Källander, N. Lampkin, S. Padel, M. Sligh, U. Niggli, G. Vogt and W. Lockeretz*	264

Index   275

Contributors

Aschemann, Jessica, Department of Agricultural and Food Marketing, Faculty of Organic Agricultural Sciences, University of Kassel, Steinstrasse 19, 37213 Witzenhausen, Germany. E-mail: j.aschemann@uni-kassel.de

Cierpka, Thomas, International Federation of Organic Agriculture Movements, Charles-de-Gaulle-Str. 5, 53113 Bonn, Germany. E-mail: t.cierpka@ifoam.org

Conford, Philip, 88 St. Pancras, Chichester, West Sussex PO19 7LR, UK. E-mail: Paconford@aol.com

DiMatteo, Katherine, Wolf & Associates, Inc., 90 George Lamb Road, Leyden, MA 01337, USA. E-mail: kdimatteo@organicspecialists.com

Foguelman, Dina, Movimiento Argentino para la Producción Orgánica (MAPO), Sarmiento 1562 6° F, Buenos Aires, Argentina. E-mail: dina5@arnet.com.ar

Fritz, Sandy, Southern Rivers Catchment Management Authority, 1194 Wattamolla Rd., Berry, NSW 2535, Australia. E-mail: Sandy.fritz@cma.nsw.gov.au

Geier, Bernward, Colabora, Alefeld 21, 53804 Much, Germany. E-mail: bub.geier@t-online.de

Gershuny, Grace, Gaia Services, 1417 Joe's Brook Road, St. Johnsbury, VT 05819, USA. E-mail: graceg@kingcon.com

Hamm, Ulrich, Department of Agricultural and Food Marketing, Faculty of Organic Agricultural Sciences, University of Kassel, Steinstrasse 19, 37213 Witzenhausen, Germany. E-mail: hamm@uni-kassel.de

Holden, Patrick, Soil Association, South Plaza, Marlborough Street, Bristol BS1 3NX, UK. E-mail: PHolden@SoilAssociation.org

Källander, Inger, Ecological Farmers Association, Gäverstad Gård, 614 94 Söderköping, Sweden. E-mail: inger.kallander@ekolantbruk.se

Lampkin, Nicolas, Organic Research Group, Institute of Rural Sciences, University of Wales, Aberystwyth SY23 3AL, UK. E-mail: nhl@aber.ac.uk

Lockeretz, William, Friedman School of Nutrition Science and Policy, Tufts University, 150 Harrison Ave., Boston, MA 02111, USA. E-mail: willie.lockeretz@tufts.edu

Naspetti, Simona, Department DIIGA, Faculty of Engineering, Polytechnic University of Marche, Via Brecce Bianche, 60131 Ancona, Italy. E-mail: simona@agrecon.univpm.it

Niggli, Urs, Research Institute of Organic Agriculture (FiBL), Ackerstrasse, 5070 Frick, Switzerland. E-mail: urs.niggli@fibl.org

Padel, Susanne, Organic Research Group, Institute of Rural Sciences, University of Wales, Aberystwyth SY23 3AL, UK. E-mail: spx@aber.ac.uk

Schmid, Otto, Research Institute for Organic Farming (FiBL), Ackerstrasse, 5070 Frick, Switzerland. E-mail: otto.schmid@fibl.org

Sligh, Michael, RAFI-USA, PO Box 640, Pittsboro, NC 27312, USA. E-mail: msligh@rafiusa.org

Stinner, Deborah H., Organic Food and Farming Education and Research Program, The Ohio State University, Ohio Agriculture Research and Development Center, 1680 Madison Ave., Wooster, OH, 44691, USA. E-mail: stinner.2@osu.edu

Vogt, Gunter, Friedrich-Naumann-Str. 91, 76187 Karlsruhe, Germany. E-mail: guntervogt@web.de

Wynen, Els, Eco Landuse Systems, 3 Ramage Place, 2615 Flynn, Canberra, Australia. E-mail: els.wynen@elspl.com.au

Zanoli, Raffaele, Department DIIGA, Faculty of Engineering, Polytechnic University of Marche, Via Brecce Bianche, 60131 Ancona, Italy. E-mail: zanoli@agrecon.univpm.it

Foreword

When I was asked to write this foreword, it provided me with a wonderful excuse to sit back, ponder, pull together my thoughts and take a serious look at this oasis of ours, organic agriculture: what we have done, how far we have come and, of course, where we are going, how we will get there, whom we will take with us and how all the work that has gone on might provide even more benefits for the future.

When does the history of this thing we now call 'organic agriculture' begin? Some say that it actually began thousands of years ago, when hunter-gatherers settled down and took up farming. After all, farmers of the remote past certainly did not have to deal with synthetic chemicals! However, true organic agriculture is practised by intent, not by default; you do not automatically become organic simply because you never used prohibited chemicals anyway. This makes it clear that organic agriculture started much more recently. It is hard to specify exactly when, but early landmarks include the founding of biodynamic agriculture in the 1920s, the emergence of a strongly organized movement in the UK in the 1940s and the promulgation of the first organic production standards in the 1960s.

Whenever it may be considered to have begun, throughout most of its history organic agriculture grew without institutional and political support, to the surprise of many sceptics. The sector is expanding not only economically (i.e. market share), but also in its sociocultural importance. We have gone from 'earth mothers' to highly respected reformers who have developed the concepts which, in turn, have led to standards that not only provide the basis for growing products while respecting the environment, but have recently also been including social justice standards.

Philosophical, environmental and food safety issues have driven the demand for organic commodities, but organic agriculture also restores a human face to agriculture by giving attention to an 'agri-culture' that values farmers' work and local traditions and foods. Somehow, the organic ethic has tried to deal with some of the overwhelming aspects of globalization by balancing science and morality.

This book provides a timely opportunity for all of us – those who have been involved with organic agriculture for a long time, as well as those who are circling on the outside trying to decide what they think of all these ideas about agriculture – to get an overview of what needs to be done in a variety of contexts to continue reaping benefits, but also to share those benefits with others. An understanding of the growth of organic agriculture in developed countries – including both its achievements and setbacks – offers newcomers from the developing world a headstart in building upon their traditional systems, with a view to creating more sustainable food systems.

Of course, organic agriculture has potential for farmers in all nations. But think about what it means in the developing world in particular, where farmers – struggling to feed themselves and their families and hoping to have food left over to sell in the market – can take advantage of a system that allows them to flourish in the absence of external support.

While modern agriculture has unilaterally privileged a scientific and economic model based on mechanistic and quantitative concepts of the non-organic world, the organic community complements scientific investigation with active creativity, whereby farmers can rely on their intelligence instead of capital and on their own knowledge and labour to add value to locally available resources. Contrast this with the global trends towards unbalanced diets (more animal products, sugars and fats), the concentration of food markets in the hands of a few large retailers and reduced national control over the flow of resources. Resource degradation from intensification and specialization poses challenges for agriculture and society as a whole.

Organic agriculture offers solutions, but to take advantage of them it is essential to have informed consumers – consumers who appreciate that organic agriculture not only promises them safer environments and foods, but also promises that the rights and traditions of producers have been respected. Furthermore, it is critical to have more ethical entrepreneurs who are willing to be transparent and share information about their products and activities. There are several examples in the organic community of individual responsibility and economic partnerships. My hope is for them to prosper – courageously – and to escape the trap of 'conventionalism'.

Yes, organic agriculture has matured enough to offer lessons. Now it is up to us to look to the future, to make sure our standards are realistic and adaptable to unique situations. Then we will be in a position to assure that our maturity does not lead to a midlife crisis. Signs of a crisis are evident today where organic systems follow the simplified production and distribution pattern of conventional agriculture in order to compete with the dominant food supply system.

Without a doubt, diverse forms of organic management will continue to be practised to suit different needs and different markets. Certain organic systems are more environmentally or socially just than others, but all have to obey the common principles of health, ecology, fairness and care, as reflected in the basic organic standards, which are constantly improved as knowledge advances. Our choices for future directions will certainly benefit from an understanding of the history of organic agriculture in the overall context of agricultural and societal development – and investing in further investigation.

This book describes the evolution of organic agriculture in the various spheres of the food supply chain. In it we can see movement between two poles: from northern to southern countries; from smallholders to industrial production; from hippie culture to scientific perspectives; from input substitution techniques to holistic approaches; from farmers' standards to government regulations; from informal markets to supermarkets; and from third-party certification to participatory guarantee systems.

In telling the story of what has happened so far, this book is inspiring. But it also makes me ponder: what next? The book concludes by offering some general possibilities, but does not propose definitive, comprehensive answers. Instead, it challenges us to be creative in speculating about what the future of organic agriculture is likely to be, or ought to be, if it is to remain a vibrant and innovative movement for cultural and social development.

Nadia El-Hage Scialabba
Senior Officer (Environment and Sustainable Development)
Food and Agriculture Organization of the United Nations
E-mail: nadia.scialabba@fao.org

This page intentionally left blank

# What Explains the Rise of Organic Farming?

W. Lockeretz

Professor, Friedman School of Nutrition Science and Policy, Tufts University, 150 Harrison Ave., Boston, Massachusetts 02111, USA

From very modest beginnings in the first half of the last century, organic farming has grown dramatically in importance and influence worldwide. A few statistics tell part of the story: from almost negligible levels until the 1980s, the number of organic farms worldwide has grown to an estimated 623,000, with some 31.5 million ha managed organically (Willer and Yussefi, 2006, ch. 5). Worldwide sales of organic products reached some US$28 billion in 2004 (IFOAM, 2006).

But these numbers capture only a small part of what organic farming has become. Even more impressive is its heightened stature among researchers, educators and agricultural policy makers, a change that began in earnest only by the late 1970s.

At one time, organic farming was fair game for attacks, with or without supporting evidence. Thus it was that at the annual meeting of the American Association for the Advancement of Science (AAAS) in 1974, a panel of scientists took aim at the 'organic food myth', calling it 'scientific nonsense' and the domain of 'food faddists and eccentrics', and blaming 'pseudoscientists' for confusing the public and thus scaring them into paying more for food (*Washington Post*, 1974). They also said that the 'organic myth was counterproductive to human welfare, because the myth leads to a rejection of procedures that are needed for the production of nutritious food at maximum efficiency' and was 'eroding gains of decades of farming advancements'. Yet, 7 years later, the journal of this same AAAS published a major research paper that found organic farms to be highly efficient and economically competitive while using less fossil energy and suffering less soil erosion than neighbouring conventional farms (Lockeretz *et al.*, 1981).

In light of the once hostile attitudes of at least some scientists (it is hard to say how prevalent these attitudes were), the growth of research on organic farming has been particularly striking. At the first international scientific conference of the International Federation of Organic Agriculture Movements (IFOAM), held in Switzerland in 1977, a total of 25 presentations was offered. When the IFOAM conference returned to Switzerland in 2000, that number had jumped by more than 20-fold, to well over 500 (Alföldi et al., 2000). Before the 1970s, funds for organic research were extremely limited; today, significant public money is available in many countries: Denmark, France, Germany, Sweden, Switzerland and the Netherlands are all reported to spend at least €6 million per year on organic research (Slabe, 2004, p. 11). Often this research also involves publicly funded advisory and outreach programmes for current and potential organic farmers, yet another activity that would have been hard to imagine in the early 1970s. Equally unimaginable are the abundance and variety of organic curricula and degrees offered at universities in many countries.

A similar evolution has been evident in the attitudes of policy makers. Back in 1971, Earl Butz, at the time the Secretary of the US Department of Agriculture (USDA), whose department had never done any research or anything else regarding organic farming, nevertheless made bold to declare that 'before we go back to an organic agriculture in this country, somebody must decide which 50 million Americans we are going to let starve or go hungry' (Butz, 1971). Yet, less than a decade later, under another Secretary of Agriculture, Bob Bergland, that same department undertook a comprehensive study of organic farming to understand better its potential and its limits, and to recommend how the USDA should get involved in it. The resulting report (USDA, 1980), which Bergland enthusiastically endorsed, was one of the most widely requested reports in USDA's history. It had a startling impact, not just because of what it said, but also because of who was saying it. The report concluded that organic farming would receive an impetus from increasing concerns over energy shortages, declining soil productivity, soil erosion, chemical residues in foods and environmental contamination. It also noted that 'the negative attitudes of...the agricultural establishment toward organic farming have sometimes limited [its] acceptance', and that the common view in the establishment that organic farming is 'impractical or infeasible' was 'to some extent...the result of misperceptions and misunderstandings' (p. 83). It offered 19 recommendations regarding organic research, education and extension, the last of which was that 'it is of utmost importance that USDA develop research and education programs and policies to assist farmers who desire to practice organic methods' (p. 93).

In retrospect, Bergland's view clearly has supplanted that of Butz in agricultural policy circles, both in the USA and in many other countries. Many governments that long ignored organic farming now offer farmers subsidies for producing organically (see Chapter 6). This is commonly done because of its environmental benefits (though a more sceptical view is that the real goal is to reduce agricultural surpluses because organic farming is thought to have lower yields). Likewise, many agencies now collect statistics on organic production and some promote it through public education campaigns and market development activities. A related development has been the proliferation of national and international regulations, standards and labels – public as well as private (see Chapter 8).

An important component of the advancement of organic farming has been its global spread. Five countries were represented when IFOAM was organized in 1972; by the late 1990s it had members from over 100 countries (see Chapter 9). IFOAM's scientific conferences, which until the mid-1980s had been held only in western Europe and North America, have since been held in countries as diverse and dispersed as Burkina Faso, Australia, Hungary and Brazil, among others. Further evidence that organic farming has become truly global is that the UN Food and Agriculture Organization has been involved in it starting in 1999, with activities that include networking, market analysis, environmental impact assessments, improving technical knowledge, responding to country requests for assistance, and development of standards through the Codex Alimentarius Commission (FAO, 2005). Similarly, the United Nations Conference on Trade and Development has been involved in several aspects of global trade in organic foods since 2001, particularly in assisting developing countries to increase their production and exports (Twarog, 2002).

This book will document many more examples showing the remarkable rise in the importance and stature of organic farming. This leads to an obvious but difficult question: What caused this rise?

One immediate answer (at least for western Europe) is the subsidization of organic farming (described in detail in Chapter 6). But subsidies, while stimulating conversion to organic farming since the 1990s, were a result, not a cause, of the earlier growing interest in organics in the 1970s and 1980s. So what caused that?

In the absence of a thorough analysis, we can only speculate. But several possibilities suggest themselves:

- Organic activists were successful in promoting their views to the public, scientists and policy makers.
- As new concerns emerged regarding the environment, the situation of farm workers and small farmers worldwide, and food safety – the last of these sometimes involving outright scandals

and near-panic – organic farming became a more attractive alternative to the dominant farming systems among both farmers and the public.
- Over the decades, organic farming changed in ways that made it more appealing to a broader public, in contrast to its narrow circle of adherents in the early days.

Impressive national and international organizations have emerged that promote organic farming among farmers, researchers, consumers and others. Some of their activities have included public education and other measures aimed directly at increasing the importance of organic farming. Other efforts have been directed towards people already involved in organic farming, such as exchanging information on improved production methods among organic farmers. But these activities no doubt also have indirect effects on the outside world, for example, when farmers who are thinking about going over to organic farming see improved methods at work on real organic farms.

But the effectiveness of such advocacy organizations should not be overstated. Their resources are far too limited to bring about so great a challenge to how we produce food. Also, some have not worked as effectively as they might have, even with their limited resources, because of internal conflicts or a reluctance to cooperate with other similarly motivated groups. Moreover, many such organizations came into existence after organic farming began its rapid growth in the early 1970s. Thus, while they might deserve credit for some recent advances, they were a product of the most striking period of growth, not its cause.

Thus, we need to look elsewhere for the rest of the explanation. The 1960s, the decade preceding the initial rapid rise in organic farming, was a time of great social and political upheaval worldwide. This markedly heightened public awareness of environmental threats, including from agriculture, and created a strong determination to do something about them. The most dramatic threat from agriculture came from pesticides, publicized so effectively by Rachel Carson (1962). One of the early successes of the environmental movement was scored in the early 1970s, when DDT and other organochlorines were banned in many countries, largely because of the harm they did to birds of prey and other threatened species. Less dramatic, but still a cause for concern, was the risk of methaemoglobinaemia ('blue baby') from the elevated nitrate levels found in drinking water supplies, in large part as a result of high applications of inorganic nitrogen fertilizers.

At the same time, a generalized rejection – or at least a greater suspicion – was developing towards synthetic chemicals of all sorts, including not just pesticides, but also food additives. This prompted growing interest in foods that were less processed and considered more wholesome, natural and safer than what was otherwise available.

The public's concern regarding the environmental and food safety implications of agricultural chemicals was paralleled by growing concern among farmers regarding their effects on their own health and that of their families and livestock. This concern was an important reason for farmers to convert to organic farming (e.g. Lockeretz and Madden, 1987).

All these developments clearly favoured the growth of organic farming, which offered more natural foods produced in safer and more environmentally sound ways. But starting in the 1960s other factors may also have boosted the organic farming movement, although less directly. The 1960s was a decade of strong anti-establishment activism, especially in opposition to the Vietnam War. Important targets of this activism were the giant petrochemical companies that made war materials such as napalm and the herbicide Agent Orange. These companies also manufactured insecticides (organochlorine and organophosphate insecticides both have their origins in World War II). Environmental and antiwar activists became natural allies in their campaigns against big chemical companies. Thus, the organic farming movement found itself very much in tune with the zeitgeist of the 1960s and early 1970s. Ironically, it gained followers on the left side of the political spectrum, whereas originally, to the extent that it was political at all, at least in England the organic movement leaned decidedly towards the right (Conford, 2001).

Finally, the 1960s are well known as a time of countercultural revolution, embodied in the stereotypic 'back-to-the-land hippie'. Today, most people involved in organic farming do not at all fit that stereotype, and are quick to point out that organic farming is *not* hippie farming. (As Danbom (1995, p. 267) described the situation in the USA: 'In 1980, most farmers viewed organic producers...with bemused contempt, as latter-day hippies with harebrained notions of how farming should be done. But their successes in the market have quieted the skeptical.') However, since the 1930s, there always have been some organic farmers and advocates who in fact could be characterized as proto-hippies, or at least as proponents of an alternative lifestyle (as discussed in Chapter 2).

This last point leads directly to the third suggested reason for the growth of organic farming, namely, changes in organic farming itself (no doubt in large measure in response to changes in the outside world). From the start, various kinds of people have favoured organic farming, for various reasons. But the relative importance of these groups has shifted over the decades.

As noted, some people find organic farming attractive because it represents a desirable alternative lifestyle, one that especially rejects the dominance of industrial power and materialistic values. This may or may not be coupled to metaphysical beliefs concerning the cosmos and life forces (critical elements in biodynamic agriculture, the first formal organic concept, as described in Chapter 2). Still others are interested in it mainly for its potential to reduce agriculture's damage to the environment. Another group, which overlaps the previous one, particularly likes organic farming as a source of wholesome, high-quality foods.

In recent years, the last two groups have become more important components of the mix, and the face of organic farming's constituency has changed. Along with the countercultural types who (among others) were there from the beginning, organic farming now is also the domain (among others) of decidedly non-hippie officials such as may be found in Brussels, promulgating EU regulations specifying the definition of 'organic' (or aging professors who teach about it in universities, such as in Boston). The process may have fuelled itself. As more 'respectable' types got involved, the movement no doubt was seen as less exotic by others who already shared its environmental goals, but not its embrace of alternative lifestyles.

Organic farming's new adherents include farmers who converted from conventional farming – perhaps inspired by the example of others who had already done so and were benefitting from the gradual but steady improvement in organic production techniques (see Chapter 5) – but who in many respects still were more like their conventional counterparts than those who had farmed organically all along. They also include perfectly respectable agricultural scientists, some of whom had distinguished careers in very conventional kinds of research before turning their attention to organic farming, possibly after seeing some of their colleagues do so as the mutual suspicions between conventionally and organically oriented scientists began to subside (see Chapter 4). The new adherents also included consumers who now could shop for organic products in otherwise conventional supermarkets, rather than having to go to health food stores that were perceived – rightly or wrongly – as just for health food 'faddists' (to use the language of the AAAS panelists mentioned earlier); again, the demand for organic products among such consumers in turn could lead more supermarkets to offer them, which in turn would increase the number of consumers buying them (as discussed in Chapter 7).

This last suggested cause of the rise in organic farming – changes in who is involved with it – has been viewed with misgivings in some circles, their fear being that newer participants will distort or dilute the fundamental principles of organic farming. Thus, conventional farmers who recently converted to organic methods are suspected of being in it just for the money (at least in countries that subsidize organic farming heavily, or where price premiums have been particularly strong). The concern is that they will do as little as possible to get certified, and possibly try to weaken the organic standards. Selling organic products through chain supermarkets (whether conventional or the more recent natural food chains) worries some people who fear that the concentrated economic power of these chains will enable them to force down the prices that organic farmers receive; so too, they fear that large supermarkets will favour larger, highly specialized and often more distant farmers who can supply the high volumes and standardized products they demand, in contrast to the earlier organic ideal of a decentralized marketing system based on small, diversified, local farms (as discussed in Chapters 3 and 7). These arguments were already heard in the 1990s, when the organic sector was considerably smaller than it is today (e.g. Woodward *et al.*, 1997; Kirschenmann, 2000; Klonsky, 2000); these days they are heard much more frequently.

A hotly debated subject in organic circles is whether supporters of organic farming should welcome the growth that comes from these changes in who is involved with it. Against the arguments just set forth, others argue that the entrance of new (and possibly larger) farmers at least means more land is being cultivated in an environmentally benign way. And whatever else one thinks of supermarkets, their growing role in organics means that more food is available that has demonstrably lower pesticide residues.

Like any good question about organic farming that is worth debating, no simple answers can, or should, be offered. So, too, with the title question of this chapter. Much more analysis is needed to construct a credible explanation of the intriguing rise of organic farming. This book does not offer an explanation. Rather, it tries to supply some building blocks from which one can be constructed.

References

Alföldi, T., Lockeretz, W. and Niggli, U. (eds) (2000) *IFOAM 2000: The World Grows Organic. Proceedings of the 13th International IFOAM Scientific Conference, 28 to 31 August 2000*. vdf Hochschulverlag AG an der ETH Zürich.

Butz, E. (1971) Quoted in *Yearbook of Spoken Opinion: What They Said in 1971*, p. 123.

Carson, R. (1962) *Silent Spring*. Houghton Mifflin, Boston, Massachusetts.

Conford, P. (2001) *The Origins of the Organic Movement*. Floris Books, Edinburgh, UK.

Danbom, D.B. (1995) *Born in the Country: A History of Rural America*. Johns Hopkins University Press, Baltimore, Maryland.

FAO (2005) Organic agriculture at FAO. United Nations Food and Agriculture Organization. Available at: www.fao.org.organicag (updated December 2005)

IFOAM (2006) The world of organic agriculture: more than 31 million hectares worldwide (press release, February 14). International Federation of Organic Agriculture Movements, Bonn, Germany. Available at:

www.ifoam.org/press/press/Statistics_2006.html

Kirschenmann, F. (2000) Organic agriculture in North America: the organic movement. In: Allard, G., David, C. and Henning, J. (eds) *L'agriculture biologique face à son développement – Les enjeux futurs.* Lyon (France) 6–8 décembre 1999. INRA Editions, Paris, pp. 63–74.

Klonsky, K. (2000) The consumption of food in the United States: confluent demand and availability. In: Allard, G., David, C. and Henning, J. (eds) *L'agriculture biologique face à son développement – Les enjeux futurs.* Lyon (France) 6–8 décembre 1999. INRA Editions, Paris, pp. 227–246.

Lockeretz, W. and Madden, P. (1987) Midwestern organic farming: a ten-year follow-up. *American Journal of Alternative Agriculture* 2, 57–63.

Lockeretz, W., Shearer, G. and Kohl, D. (1981) Organic farming in the Corn Belt. *Science* 211, 540–547.

Slabe, A. (2004) *Consolidated Report: Second Seminar on Organic Food and Farming Research in Europe: How to Improve Trans-national Cooperation.* Available at: http://www.agronavigator.cz/attachments/CORE_seminar_listopad_2004.pdf

Twarog, S. (2002) UNCTAD's work on organic agriculture. In: Rundgren, G. and Lockeretz, W. (eds) *Reader, IFOAM Conference on Organic Guarantee Systems: International Harmonization and Equivalence in Organic Agriculture,* 17–19 February 2002, Nuremberg, Germany. IFOAM, Tholey-Theley, Germany, p. 51.

USDA (1980) *Report and Recommendations on Organic Farming.* United States Department of Agriculture, Washington, DC.

Washington Post (1974) Organic farming 'scientific nonsense'. 28 February.

Willer, H. and Yussefi, M. (eds) (2006) *The World of Organic Agriculture: Statistics and Emerging Trends 2006.* International Federation of Organic Agriculture Movements, Bonn, Germany. Available at: www.orgprints.org/5161

Woodward, L., Fleming, D. and Vogtmann, H. (1997) Health, sustainability and the global economy: the global dilemma. *Ecology and Farming* 17, 32–39.

2 The Origins of Organic Farming

G. Vogt

Friedrich-Naumann-Str. 91, 76187 Karlsruhe, Germany

2.1 Introduction

The concept we know today as 'organic farming' is an amalgam of different ideas rooted mainly in the German-speaking and English-speaking worlds. These ideas arose at the end of the 19th century, especially the knowledge of biologically oriented agricultural science, the visions of Reform movements and an interest in farming systems of the Far East.

Between the two World Wars 'modern', chemical-intensive, technically advanced farming faced a crisis in the form of soil degradation, poor food quality and the decay of rural social life and traditions. As a solution to this crisis, organic farming pioneers offered a convincing, science-based theory during the 1920s and 1930s that became a successful farming system during the 1930s and 1940s. But it was not until the 1970s, with growing awareness of an environmental crisis, that organic farming attracted interest in the wider worlds of agriculture, society and politics.

The leading strategies proposed to achieve sustainable land use included a biological concept of soil fertility, intensification of farming by biological and ecological innovations, renunciation of artificial fertilizers and synthetic pesticides to improve food quality and the environment and, finally, concepts of appropriate animal husbandry.

2.2 Context of the Origins

Organic farming developed almost independently in German-speaking and English-speaking countries in the early 20th century. Its origins need to be understood in the context of four developments going on at the time: (i) a crisis in agriculture and agricultural science; (ii) the emergence of biologically oriented agricultural science; (iii) the Life and Food Reform movements; and (iv) growing Western awareness of farming cultures of the Far East.

2.2.1 Crisis in agriculture and agricultural science

Agriculture and agricultural science underwent a crisis between the two World Wars in which they faced ecological and soil-related as well as economic and social problems. The use of mineral fertilizers, pesticides and machinery – the chemical-technical intensification of farming – was variously seen as either a cause of or a solution to these problems.

Scientific and agricultural debates in Germany discussed the increased use of mineral fertilizers and the corresponding neglect of organic manuring (summarized in Vogt, 2000a) as a major cause of several problems:

- Inappropriate use of mineral fertilizers was disturbing plant metabolism, especially because cultivars at that time were not yet adapted to higher nitrogen levels in soil. Weakened plants could be attacked more easily by pathogens and insect pests, and effective pesticides had not yet been developed.
- Physiologically acidic mineral fertilizers acidified the soil, leading to diminished root growth, disturbances in the soil's mineral balance and degradation of soil structure.
- Soil compaction caused by the use of machinery and reduced organic manuring lowered the soil's water-retaining capacity, causing drought problems.
- Soils experienced a decline in fertility – referred to as 'soil fatigue' (*Bodenmüdigkeit*) – that could not be explained by harmful organisms or the lack of nutrients; this was attributed to a disturbed balance among soil organisms, with the resulting accumulation of harmful organic substances.
- The use of the previous harvest as seeds often led to a decrease in yields that could not be explained by plant diseases, pests or mineral deficiencies. Higher nitrogen levels in soil and plants prevented the complete ripening of the seeds; such immature seeds interfered with the plant's development the following year.

Similar discussions regarding decreasing soil fertility had arisen in the UK and the USA with different starting points, respectively management problems and concurrent yield decreases, and the Dust Bowl in the Great Plains.

Despite an increased use of mineral fertilizers, German agriculture suffered from a dramatic drop in yields (up to 40%) after World War I; only at the end of the 1930s – after more than 15 years – did yields again reach pre-war levels (Bittermann, 1956). These yield decreases were at least partly linked to the above-mentioned problems, which were attributed to the increased use of mineral fertilizers in combination with the lack of suitable cultivars and pesticides, as well as by the neglect of organic fertilization. At the time, some German agricultural scientists did not believe in the long-term success of mineral fertilizers and feared an overexploitation of soil fertility; they attributed the yield losses to the short-term success of mineral fertilizers before World War I (Vogt, 2000a).

In addition, some consumers were worried about declining food quality: food that did not stay fresh, tasteless vegetables and fruits, and residues from pesticides based on toxic elements such as arsenic, mercury or copper. The public discussed the increased use of mineral fertilizers and pesticides as a major cause of this decline, and suspected, for example, that an elevated level of potassium in cancer cells was caused by increased potassium fertilization. Scientists like Robert McCarrison in the UK or Werner Schuphan and Johannes Görbing in Germany confirmed some of these suspicions, such as lower vitamin levels in fruits and vegetables caused by increased nitrogen fertilization (McCarrison and Viswanath, 1926; Schuphan, 1937).

Finally, the social and economic situation in the countryside changed dramatically with the mechanization of agriculture, industrialization of the food sector, migration from the land and import of agricultural products. An imbalance arose between the urban centres and the countryside, and national food self-sufficiency no longer was guaranteed. Severe economic problems caused by low prices (due to imports) and indebtedness (due to purchase of machines, fertilizers and pesticides) forced many small and medium-sized farms to give up. Furthermore, social life in the countryside saw a decline of rural tradition and rural lifestyle.

2.2.2 Biologically oriented agricultural sciences

Following the discovery of mycorrhizal fungi by Albert Bernhard Frank in 1885 and nitrogen-fixing bacteria by Hermann Hellriegel and Hermann Wilfahrt in 1886, soil biologists started to investigate the soil from a biological point of view. At the turn of the century a new agricultural discipline emerged: agricultural bacteriology, dealing with bacteria in soil, manure, silage and milk. Soil biology pioneers included Felix Löhnis (1874–1931), Lorenz Hiltner (1862–1923) and Raoul Heinrich Francé

(1874–1943) in Germany, and Selman A. Waksman (1888–1973) in the USA. In 1910, Löhnis' *Handbuch der landwirtschaftlichen Bakteriologie* (Handbook of Agricultural Bacteriology) was published, the first definitive book on soil biology (Löhnis, 1910).

By integrating the research findings of these pioneers, agricultural bacteriology developed an inclusive biological concept of soil fertility focusing on the community of soil organisms, the dynamics of soil organic matter and the relations between plant roots and soil. This concept of soil fertility recommended feeding the soil organisms by organic fertilization (rotted organic material and green manuring), whereas the agrochemical approach recommended the increase of soil minerals. By applying those biological research findings to farming practice, scientifically trained farmers improved farming methods in areas such as soil cultivation, composting, organic fertilization, green manuring and crop rotation. In this scientific point of view, organic farming is an intensification of farming by biological and ecological means in contrast to chemical intensification by mineral fertilizers and synthetic pesticides.

Agricultural bacteriology called the dominant agrochemical theories into question. Subsequent fundamental debates regarding the importance of soil organisms and organic fertilization arose between agricultural biologists and chemists in Germany during the 1920s, 1930s and 1940s. Agricultural biologists stated that organic manure is the prerequisite for (small) additional mineral fertilization; they emphasized biological and chemical interactions, such as biological nitrogen fixation, active mobilization of minerals by soil organisms and plant roots, as well as soil organic matter as reservoirs of minerals. Agricultural chemists made a strict distinction between mineral fertilization and organic manuring: the (more important) mineral fertilization to increase the soil's mineral content (and yields), and organic manuring to sustain the soil's biological fertility. Only one interaction was mentioned: mineral fertilization leading to higher yields will produce more organic plant residue, which increases soil organic matter (Vogt, 2000a).

2.2.3 Life Reform and food reform movements

Starting at the end of the 19th century, reform movements such as the German 'Life Reform' (*Lebensreform*) and the American 'Food Reform' disapproved of industrialization, urbanization and the growing dominance of technology in the 'modern' world. They called for a 'natural way of living' consisting of vegetarian diets, physical training, natural medicine and going back to the land. Other interests included abstinence from alcohol and other drugs; educational reform; protection of nature, animals, and local and regional native culture; and garden cities and garden plots.

The movements consisted of a variety of associations that were continually being founded and disbanded. To disseminate their ideas they gave public lectures and published countless journals covering all topics of the movements. Reform stores, vegetarian restaurants and natural nursing homes offered products and services necessary for a 'natural' way of living. Finally, they organized expositions on Life Reform topics and took part in official hygiene expositions (Krabbe, 1974).

Eating habits changed dramatically with the increases in industrial food processing and high-meat diets that were rich in fats and protein but poor in fibre. Diseases caused by 'modern' nutrition appeared or increased: overweight, indigestion, circulation disorders, diabetes and caries. Food Reform proposed a vegetarian or low-meat diet with little or no industrial food processing; this diet was thought comparable to the supposed diet of early human beings.

Thus, Life Reform and organic farming met on two points: going back to the land and farming organically on the one hand, and nutrition through healthy, organically grown food on the other. But organic farming did not become a key part of the urban Life Reform movement: vegetarian nutrition played a more important role than high-quality organic food, and only a few members of the Life Reform movement dared to leave the urban centres, settle on the land and work as farmers (Baumgartner, 1992).

2.2.4 Farming cultures of the Far East

People involved in the early development of organic farming admired the farming cultures of the Far East because of their sustainability over centuries and millennia, and many aimed at transferring Far Eastern farming concepts to European agriculture. They were influenced by reports on voyages to Far Eastern countries that focused on agriculture, such as that of Franklin H. King in the early 20th century (King, 1911). Organic farming tried to adopt several farming and gardening practices, including composting techniques, transplanting cereals (such as wetland rice) and recycling of municipal organic waste.

In retrospect, however, Far Eastern farming systems had almost no practical influence on organic farming, in several respects:

- From the beginning organic farming preferred aerobic composting instead of the anaerobic methods used in the Far East.
- Efforts to mimic wetland rice farming – starting cereal plants in small beds and then transplanting the young plants to large fields – failed.

- Only a few local projects in the 1950s and 1960s attempted to recycle municipal organic waste from households, factories and sewage treatment plants to use them as fertilizers. Nowadays fertilization with sewage sludge is forbidden for organic farmers because of contamination by heavy metals and other harmful substances.
- The replacement of water closets by composting toilets, a goal during organic farming's pioneer period, was never achieved.
- A vegetarian diet was not in accordance with Far Eastern habits.

Nevertheless, the Far East played a key role in the development of organic farming by presenting a model of a sustainable society based on gardening and farming.

2.3 Natural Agriculture and Its Successors in the German-speaking World

Two main currents of organic farming had been established in German-speaking countries by the early 20th century: the science-based natural agriculture (which, being a part of the Life Reform movement, also called itself 'Land Reform') and the anthroposophic biodynamic agriculture (since 1924).

2.3.1 Concepts of natural agriculture

Since the beginning of the 20th century the Life Reform movement proposed a 'natural way of living'. While the majority of people in the mainly urban reform movement only talked about going back to the land, some tried to realize their ideals by leaving the urban centres, living in rural nature and working as farmers and gardeners. Their concepts of organic farming included a healthful vegetarian diet, gardening, fruit growing and farming without animals.

Because of their vegetarian beliefs and their rejection of technology they had to deal with two dilemmas: their ideology was opposed not just to agricultural machines but also to draught animals, and not just to artificial mineral fertilizers but also to animal manure. Compromises regarding their beliefs, an emphasis on gardening and finally the biological understanding of soil fertility solved or at least reduced these problems.

The biological concept of soil fertility confirmed the seriousness of the farming concept they developed. Natural agriculture's soil cultivation included careful composting, conservation tillage, green manuring, rock powder fertilization and mulching. Friedrich Glanz (1922) and

Heinrich Hopf (1935) promoted conservation tillage; they recommended decompacting soil without using a mouldboard plough, which destroys the 'natural' soil layers. Heinrich Krantz (1922) developed a new composting method called 'noble manure' (*Edelmist*), combining a short period of aerobic rotting with a long period of anaerobic fermentation. Johannes Schomerus (1931) favoured a permanent soil cover of living plants or organic residues to protect the soil from drought, rain and erosion.

In their view, artificial fertilizers were responsible for decreases in food quality, soil fertility and plant health; consequently, they relied solely on organic manure. But at the same time they rejected animal manure because they could not reconcile animal husbandry with their vegetarian ideology. Therefore, green manuring and composting of plant residues had to play the key role in fertilization. They also tried to establish a municipal waste and humus economy including recycling of human faeces by composting toilets. Composted urban wastes were to be used as organic and mineral fertilizers to replace lost minerals. Finally, rock powder was used as mineral fertilizer to replace minerals removed in the harvest.

Natural agriculture's adherents faced another dilemma regarding farm work: they refused to use draught animals because of their vegetarian beliefs, or to adopt agricultural machines, which were considered incompatible with a natural way of living. However, rural everyday life forced them to use agricultural machines; consequently, they sought to develop small-scale and intermediate agricultural technology suited to organic farming and gardening. They also accepted a small number of animals. As a result, they developed the first concepts of appropriate animal husbandry: high-quality fodder, grazing on pastures and a high standard of hygiene.

2.3.2 Organization and pioneers

The Arbeitsgemeinschaft Natürlicher Landbau und Siedlung (Natural Farming and Back-to-the-Land Association) was founded in 1927/28. It developed the first standards for organic farming, which were published in the movement's monthly journal *Bebauet die Erde* (which translates as both 'Cultivate the Soil' and 'Cultivate the Earth') in 1928 and 1933. Organic products produced according to those standards could be sold under the trademark 'Biologisches Werterzeugnis' (Biological Premium Product), which was displayed with different designs in 1933 and 1937. Training and advisory projects complemented their activities. The Siedlerschule Oberellen (Oberellen Back-to-the-Land School) offered several courses in 1933, lasting from a weekend to half a year. In December 1934, forced into line by the Nazi government, the association had to join – as Arbeitsgemeinschaft Landreform (Land Reform Association) – the Deutsche Gesellschaft für Lebensreform (German Society for Life Reform).

The journal *Bebauet die Erde* was founded by Walter Rudolph in 1925. It reported on the movement's efforts, failures and successes. The main focus of the journal was to promote sustainable soil cultivation based on biological grounds. The journal was the main forum to exchange experiences by means of articles and readers' letters. Classified advertisement had been an important marketplace for seeds, (small) machines, work, land and finally organically grown food. From the 1930s it was edited by Ewald Könemann and Erich Siebeneicher until it was closed down in 1943 because of World War II.

The key person of natural agriculture was Ewald Könemann (1899–1976): 'by plough and book' he combined the various biologically based farming methods into a convincing, scientifically based organic farming concept. His first article on organic farming, 'Viehloser Ackerbau – naturgemäße Bodenbearbeitung' (Farming without Animals – Natural Soil Cultivation), was published in the Life Reform journal *TAO* in 1925 (Könemann, 1925). His three-volume work *Biologische-Bodenkultur und Düngewirtschaft* (Biological Soil Culture and Manure Economy), published in 1931, 1932 and 1937 (Könemann, 1939), summed up the principles of natural agriculture. Besides countless articles in *Bebauet die Erde* as well as in other Life Reform and agricultural journals, Könemann published several brochures on practical issues such as composting, manuring, plant protection and food preservation. He was also engaged in organization, training and marketing.

Before Könemann, several members of the Life Reform movement published instructions for organic gardening aimed at healthy nutrition, self-sufficiency and going back to the land. First came Gustav Simons' *Bodendüngung – Pflanzenwachstum – Menschengesundheit* (Soil Fertilization – Plant Growth – Human Health) (Simons, 1911). Starting in 1925, Richard Bloeck published many articles on soil fertility and soil cultivation in *Bebauet die Erde*. Other important books were *Der natürliche Landbau als Grundlage des natürlichen Lebens* (Natural Farming as the Basis for Natural Living) (Rudolph, 1925) and *Bodenfruchtbarkeit durch neuzeitliche Bodenbearbeitung* (Soil Fertility by Modern Soil Cultivation) (Herr, 1927). Wilhelm Büsselberg's *Natürlicher Landbau – Bodenständige und gesunde Ernährung* (Natural Agriculture – Native and Healthy Nutrition) (Büsselberg, 1937) was published in 1937 but was banned by the Nazi authorities before being distributed.

In Switzerland, Mina Hofstetter (1883–1967) picked up the ideas of Ewald Könemann. Her farm at Ebmatingen near Zurich was an experiment station as well as a training centre. She wrote several books on organic farming (Hofstetter, 1942), published regularly in Swiss and German Life Reform journals and gave lectures all over Europe. Other Swiss pioneers included Anna Martens and Hans Schwager in the field of organic fruit growing and gardening (Martens and Schwager, 1933). Finally, Wilhelm and Karl Utermöhlen experimented with rock powder fertilization inspired by the book *Brot aus Steinen* (Bread from Stone) (Hensel, 1939) written by the Life Reform doctor Julius Hensel in 1898.

2.3.3 Further development: natural, biological and ecological agriculture

The science-based tradition of organic farming in Germany and Switzerland continued after World War II. During the 1950s and 1960s organic farming was called biological agriculture, and later during the 1980s and 1990s, ecological agriculture. Most important, some key principles of Life Reform had been abandoned: vegetarianism, farming without animals, back-to-the-land concept and recycling of municipal organic wastes. Without these principles organic farming came closer to the mainstream of agriculture, society and politics (Vogt, 2000a).

Organic farming's proponents incorporated present-day knowledge of the biologically oriented agricultural sciences: biologically stabilized soil structure (*Lebendverbauung*), rhizosphere dynamics and systems ecology. They also developed ecological technologies concerning soil management, plant cultivation and appropriate animal husbandry. Important innovations had been Johannes Görbing's spade diagnosis (Görbing, 1947) and Richard Köhler's concept of 'bio-technical' farming (Köhler, 1949). Ernst Weichel invented several soil cultivation tools to decompact soil and recommended the use of combinations of tools to minimize soil-compacting operations.

During the 1950s organic farming, under the name 'agriculture biologique', gained a foothold in France, influenced by British and German science-based organic farming. A key figure was Claude Aubert, whose *L'agriculture biologique* (Aubert, 1970) became a fundamental book for organic farming. The French association Nature et Progrès was founded in 1964. It played a key role in the founding of the International Federation of Organic Agriculture Movements (IFOAM) in 1972 (see Chapter 9, this volume). A distinctive French approach was the 'méthode Lemaire–Boucher'; an organization of the same name was founded in 1963. Raoul Lemaire and Jean Boucher introduced the use of calcified algae as organic fertilizer; their concepts included a transformation of chemical elements proposed by C. Louis Kevran.

2.3.4 Organic-biological agriculture

In the 1950s and 1960s the Schweizerische Bauern-Heimatbewegung (Swiss Farmers' Movement for a Native Rural Culture) searched for an alternative to the industrialization of farming. The aim was to save a rural way of living rooted in Christian faith in the modern world. Led by Hans Müller (1891–1988) and especially by his wife Maria Müller (1899–1969), the farmers developed an original organic farming practice called 'organic-biological agriculture', which was characterized by ley farming, sheet composting and conservation tillage. They combined their own traditional techniques with natural agriculture, British organic farming and some experiences of biodynamic agriculture (discussed below).

The third key person of organic-biological agriculture was the German doctor and microbiologist Hans Peter Rusch (1906–1977); his concept of nature as a cycle of living particles (*Kreislauf lebendiger Substanz*) built the theoretical background of organic-biological agriculture (Rusch, 1955, 1968). He declared the existence of eternal biological entities, 'living particles'; their totality represented an ultimate 'cycle of nature'. Those living particles were able to switch between a healthy and an ill state; the fertility of soil, the quality of food and the health of organisms depend on the number of healthy living particles. He associated these entities with the DNA-containing particles of the cell; even in the early 1950s, when DNA molecules were discovered, his theory was quite strange. Based on his concept he introduced a biological soil test to indicate the quantity and quality of living particles in soil.

The Swiss journal *Kultur und Politik* (Culture and Politics) reported on organic farming starting in 1946; farmers wrote on their experiences, efforts and failures. Maria Müller reported on her experiences in organic gardening and healthy nutrition; Hans Peter Rusch regularly presented aspects of his 'cycle of living particles' and his soil test. Another key topic of the journal was rural culture and Christian faith. Organically grown food was marketed by the cooperative Heimat (Rural Home). They sold organic food to customers directly by post, to the cooperative Migros and to food enterprises belonging to the Reform movement. During the 1960s organic-biological farming concepts spread from Switzerland to Austria and Germany.

Organic-biological agriculture abandoned Rusch's concept of the 'cycle of living particles' during the 1970s, and adopted the science-based concepts of natural and biological agriculture. The two merged to become today's organic farming, called ecological agriculture in Germany. Professional organizations concerned with extension, certification and marketing were established during the 1980s. Organic farming's goals regarding agricultural and social politics changed from preservation of rural life during the 1950s and 1960s to environmental protection during the 1980s and 1990s.

2.4 Biodynamic Agriculture

The second major source of organic farming in the German-speaking world was the *Landwirtschaftlicher Kurs* (Agricultural Lectures), given by Rudolf Steiner (1861–1925) at Koberwitz near Breslau, Silesia, in 1924. An audience of about 60 persons, mainly anthroposophic farmers, listened to the eight lectures subtitled *Geisteswissenschaftliche Grundlagen zum Gedeihen der Landwirtschaft* (Spiritual foundation for the renewal of agriculture) (Steiner, 1985). Steiner did not present a complete organic farming concept; he only proposed some guidelines. Based on his outline, biodynamic agriculture was developed by a group of anthroposophic farmers. His Agricultural Lectures were first published in 1963; until then only a limited number of copies were circulated in anthroposophic circles, numbered and marked 'for personal use only' (Koepf and von Plato, 2001).

2.4.1 Biodynamic concepts

The concepts of biodynamic agriculture are derived from anthroposophy, an esoteric-occult world view. Nature is conceived as a 'spiritual–physical matrix', consisting of four levels: physical, ethereal, astral and ego forces (*Ich-hafte Kräfte*). This spiritual–physical matrix could be manipulated on the level of 'ethereal and astral forces' by the 'biodynamic preparations'. The key concept presented by Steiner is the farm as a living organism and individuality, characterized by 'ego forces'. Finally, Steiner called for an intimate 'personal relation' to nature as the basis of agricultural work (Steiner, 1985). The term 'biodynamic' was coined in 1925 by Erhard Bartsch (1895–1960) and Ernst Stegemann (1882–1943), combining two main aspects: the biological character of fertilization on the one hand, and the dynamic effects of the natural forces on the other (Koepf and von Plato, 2001).

Anthroposophists do not merely compare a farm to an organism; to them a farm *is* a real (living) organism and individuality, and like a human being it can be characterized by physical, ethereal, astral and ego forces derived from the biodynamic concept of nature. A farm organism must consist of a variety of 'organs' such as crop production, animal husbandry, gardening and fruit growing, with a diversity of plants, animals and biotopes. Based on interactions among its 'organs', as well as its adaptation to local environmental conditions, a biodynamic farm should be able to reproduce itself without supplies from outside. Finally, only a closed farm organism will attain high levels of soil fertility, plant and animal health, and food quality (Remer, 1954; Schaumann, 1994).

Although biodynamic agriculture is a successful version of organic farming, none of its four essential aspects – its concept of nature, its characteristic preparations (see below), the notion of a farm as a living organism and individuality, and the intimate, 'personal relation' to nature – have been incorporated into 'modern', science-based organic farming.

2.4.2 Biodynamic pioneers and their activities

The development of biodynamic agriculture occurred mainly on estates in the eastern parts of pre-World War II Germany. Famous biodynamic estates were Marienstein near Göttingen, Heynitz and Wunschwitz near Meissen, Pilgrimshain in Silesia and Marienhöhe in Bad Saarow. The vast estates offered favourable financial and working conditions to explore and develop a new organic farming practice. During the 1920s and 1930s, biodynamic pioneers successfully established an organic farming practice by combining the suggestions of Rudolf Steiner with traditional and modern farming methods.

The gardener Max Karl Schwarz (1895–1963) introduced elaborate composting techniques to biodynamic farming. Immanuel Vögele (1897–1959) worked on manuring. He favoured green manuring – despite the critical remarks by Steiner – and the use of composted urban organic waste. Ernst Stegemann and Immanuel Vögele were engaged in breeding cultivars adapted to biodynamic farming conditions. Ironically, the key concept presented by Steiner – the farm as a living organism and individuality – did not play any role during the pioneer period. Furthermore, while focusing solely on biodynamic preparations and manuring, there was no attempt to create concepts of appropriate animal husbandry.

A major area of emphasis was the testing of the biodynamic preparations: the field preparation (horn manure and horn silicea) and the compost preparations (yarrow blossoms, camomile blossoms, stinging nettle, oak bark, dandelion flowers and valerian flowers). Inconsistent results led to a different use of the preparations depending on the plant species, soil and climate.

From the late 1920s biodynamic agriculture was the topic of public agricultural and scientific debates on food quality, the sustainability of farming and the effectiveness of the preparations. Most field trials and farm comparisons showed lower yields on biodynamic farms, but some confirmed a higher quality of the food (summarized in Vogt, 2000a). Almost no scientific experiments specifically showed any effects of biodynamic preparations regarding plant development, yield or quality, and that is still true. Most comparative trials or farmers' field observations could not prove that biodynamic preparations had any effects because besides the use or non-use of the preparations they also included other treatment differences, such as mineral versus organic fertilization. Therefore, even if different outcomes were observed, they could not definitively be related to biodynamic preparations. In addition, the design of many comparisons either favoured or discriminated against the biodynamic treatment, depending on the intentions of the researcher or farmer. In all, there are no convincing results – from either formal research or farm observations – on the effects of biodynamic preparations that take into account different locations and last through several growing seasons (summarized in Vogt, 2000a).

A further difficulty in testing biodynamic preparations is that opinions on how they interact with nature have changed over the decades. At first, during the pioneer period, biodynamic preparations were believed to directly benefit plant development, quality, health and yield. Later (after World War II) it was stated that the supposed beneficial effects on plant development and soil fertility depended on careful organic soil cultivation. Today the effects of the preparations are not discussed any more on the level of plants and soils, but rather on the level of the whole farm organism; in addition, the preparations are said to have a regulating or normalizing effect, so that they also could lower yield, decrease quality or hinder plant development (Spiess, 1978).

2.4.3 Organizations and activities

The early biodynamic organizations consisted of the initial experimental group Versuchsring anthroposophischer Landwirte (Experiment Circle of Anthroposophic Farmers), regional associations, centres for information and advice, marketing cooperatives and a supporting society. In 1933 all these organizations merged into the umbrella organization Reichsverband für biologisch-dynamische Wirtschaftsweise (Association of the Reich for Biodynamic Farming), led by Erhard Bartsch and Franz Dreidax (1892–1964). The association joined the Nazi organization Deutsche Gesellschaft für Lebensreform (German Society for Life Reform) in 1935.

During the late 1920s fewer than 100 farms were worked biodynamically; in the 1930s estimates of the number of biodynamic farms differ widely, from no more than 200 to as many as 2000. The trademark 'Demeter' was introduced in 1928; the first standards published in 1928 distinguished between biodynamic products 'Demeter I' and conversion products 'Demeter II'. Annual conferences on biodynamic farming were held first in Berlin and Basel, and later in Göttingen and Bad Saarow (Koepf and von Plato, 2001).

In addition to several internal newsletters, the biodynamic journal *Demeter* was published starting in 1930. The journal's main topics were reports on biodynamic farms, results of field trials and debates on theoretical and practical issues. The first detailed publication on biodynamic farming was the 1929 edition of the anthroposophic yearbook *Gäa Sophia*, which was devoted to agriculture (Wachsmuth, 1929). Between 1939 and 1941 five brochures on different biodynamic farming topics were published: an introduction to biodynamic farming, by Franz Dreidax; medical plants, by Franz Lippert; fruit growing, by Max Karl Schwarz; animal husbandry, by Nicolaus Remer; and manuring, by Hellmut Bartsch and Franz Dreidax. Two books that were more theoretical were *Die Fruchtbarkeit der Erde, ihre Erhaltung und Erneuerung* (Soil Fertility: Renewal and Preservation) (Pfeiffer, 1938) and *The Agriculture of Tomorrow* (Kolisko and Kolisko, 1939).

2.4.4 Biodynamic agriculture during the Third Reich

Although the Nazi leadership disapproved of anthroposophy because of incompatible ideologies, the biodynamic umbrella organization Reichsverband für biologisch-dynamische Wirtschaftsweise was not closed down. Several Nazi leaders – Rudolf Hess, Richard Walther Darré (Minister of Agriculture and Peasant Leader of the Reich) and Heinrich Himmler – were interested in biodynamic agriculture and demanded that its potential be tested. Sceptical regarding the long-term success of artificial fertilizers, they aimed at developing a non-anthroposophic, science-based 'agriculture in accordance with the laws of life' (*lebensgesetzlicher Landbau*). Their commitment was based on various interests: sustainability of farming, food quality and soil fertility, the farm's and society's self-sufficiency regarding fertilizers, and personal esoteric interests. Nazi officials regularly visited biodynamic farms and conferences, and demanded several expert reports on biodynamic agriculture.

Alwin Seifert (1890–1972), a non-anthroposophic landscape architect and later Reichslandschaftsanwalt (Landscape Counsel of the Reich), mediated between the Nazi authorities and the biodynamic organizations. To avoid an impending ban and to spread biodynamic farming, the biodynamic leadership presented an organic farming practice without the anthroposophic background, integrated elements of the Nazi doctrine of *Blut und Boden* (blood and soil) into biodynamic concepts and willingly collaborated with the Nazi authorities. The anthroposophic gardener Franz Lippert (1901–1949) supervised the biodynamically cultivated herbal plantation at the Dachau concentration camp; several members of biodynamic organizations worked on estates of the SS; Erhard Bartsch agreed to train settlers to cultivate conquered land in the East.

This 'alliance' was not based on ideology; rather it was a tacit agreement. The biodynamic organizations were allowed to continue their work; the Nazi circles might have acquired – if biodynamic farming proved successful – results that were in accordance with their ideology. In spite of the biodynamic concessions to the Nazi authorities, the Sicherheitsdienst (Security Service), led by Heinrich Himmler, prohibited the remaining anthroposophic and biodynamic associations during a campaign against 'esoteric doctrines and lores' in June 1941. Ironically, that same month Himmler arranged to establish field experiments comparing organic and conventional farming as well as testing biodynamic preparations (Vogt, 2000b).

2.4.5 After World War II

The centres of biodynamic agriculture – the vast estates in the eastern parts of pre-World War II Germany – were lost after 1945; West Germany's family farms required a change of biodynamic farming practice and concepts. Biodynamic agriculture, like organic-biological agriculture, now focused on the preservation of rural life. The Forschungsring für Biologisch-Dynamische Wirtschaftsweise (Research Circle for Biodynamic Farming), founded in 1946, was led by Hans Heinze (1899–1997). The journal *Lebendige Erde* (Living Soil or Living Earth) first appeared in 1950.

Scientific-biological knowledge was integrated into biodynamic concepts during the 1950s and 1960s, especially by Nicolaus Remer (1906–2001), thereby bringing them closer to those of general organic farming. Simultaneously the anthroposophic aspects of biodynamic farming became less important. This 'scientificization' of biodynamic farming led to a biodynamic counter-movement initiated by Hellmut Finsterlin (1916–1989), emphasizing the esoteric-occult tradition of anthroposophy and biodynamic farming. Their journal, which appeared from 1975 to 1991, was named *Erde und Kosmos* (Earth [or Soil] and Cosmos).

During the 1980s and 1990s the focus of biodynamic agriculture passed from the preservation of rural traditions to environmental protection and sustainable farming. Concepts of appropriate animal husbandry were developed, and efforts were initiated to breed cultivars adapted to organic farming conditions. First Nicolaus Remer and later Manfred Klett and Wolfgang Schaumann explored the key concept of biodynamic agriculture, the farm as a living organism and individuality (as described above), as intended by Rudolf Steiner in his Agricultural Lectures. Many biodynamic projects successfully combine agricultural work with social work, integrating people who are handicapped or have mental, drugs-related or educational problems.

2.5 Organic Farming in the English-speaking World

The roots of organic farming in the English-speaking world can be found in India, where two scientists had been working: an agricultural scientist, Albert Howard (1873–1947), and a doctor, Robert McCarrison (1878–1960).

2.5.1 Beginnings in India

In Pusa, New Delhi, India, Howard worked on plant breeding and plant protection. At the agricultural research station at Indore, India, he developed an aerobic composting technique known as the 'Indore Process' (Howard, 1933, 1935). Another goal was to compost urban organic residues and use them to maintain soil fertility (Conford, 1995). Howard worked together with two sisters: Gabrielle Howard (1876–1930), his first wife, and Louise Howard (1880–1969), whom he married after Gabrielle's death. Both women's contributions to the development of organic farming are still underestimated. North American, as well as British, organic farming was fundamentally influenced by their teamwork (Inhetveen, 1998).

Having worked in several agricultural areas – plant breeding, plant protection, soil science, composting, manuring – Howard finally started to examine the whole farm. By reintegrating the different agricultural research disciplines, he concluded that the health of soil, plants, animals and humans are interrelated. A humus-rich soil is the key for successful (organic) farming; soil fertility is the precondition for healthy plants and animals. His famous book *An Agricultural Testament* (Howard, 1940) summarizes his experiences, emphasizing the whole farm as the starting point and basic unit of agricultural research.

Robert McCarrison's research at the Nutrition Research Laboratories in Coonoor, India, had been on the relationships among soil fertility, food quality and human nutrition. Studying the health and physique of the Hunza tribesmen living at India's north-west frontier, he discovered the meaning of nutrition for health: their nearly vegetarian diet consisted mainly of whole grains, vegetables, fruits and milk products; meat and alcohol did not play a major role. His findings opened a new perspective for medicine: to examine the conditions that determine one's health, rather than simply to cure diseases.

McCarrison also examined the decrease in food quality caused by increased use of mineral nitrogen fertilizers (McCarrison and Viswanath, 1926; McCarrison, 1936). According to Schuphan (1937), his experiments were the first to examine the relations among artificial mineral fertilizers, quality of food and human nutrition. By his observations and experiments he defined the 'Wheel of Health', consisting of soil, plants, animals and humans: properly composted organic residues will create a fertile soil, on which strong plants will grow, offering a healthy diet for humans and animals.

2.5.2 Organic farming in the UK

Influenced by Howard's concepts, farmer and animal breeder Friend Sykes (1888–1965) and herbalist Newman Turner (1913–1964) developed organic farming concepts similar to those developed in Germany. Based on a biological understanding of soil fertility they developed an organic soil management concept emphasizing ploughless soil cultivation, organic soil cover, green manuring and ley farming (Turner, 1951; Sykes, 1959). George Stapledon (1882–1960) worked in grassland cultivation: his fields of activity were the establishment and cultivation of a diverse grass turf, breeding of grassland plants and improvement of the quality of fodder (Moore-Colyer, 1999).

Inspired by the ideas of Howard and McCarrison, in the 1940s Eve Balfour (1898–1990) founded the British organic farming organization The Soil Association (described in detail in Chapter 10, this volume) and the journal *Mother Earth* (Balfour, 1943). She also initiated the Haughley Experiment, in which the effects of organic and conventional farming systems were compared at the whole farm level for some three decades. The Haughley Experiment was the first long-term experiment on organic farming.

In the 1930s the right-wing journal *New English Weekly – Review of Public Affairs, Literature and the Arts* sympathized with the beginnings of the organic farming movement. The journal worried about a disturbed balance between urban centres and rural land, a lost national self-sufficiency in food, and the vanishing of small and medium-sized farms. They also put ecological issues on their agenda, such as decreasing soil fertility and food quality. Organic farming was advocated as a solution to these rural (and urban) problems. Many pioneers of British organic farming published in the *New English Weekly* (Conford, 2001).

2.5.3 Organic farming in the USA

Since the beginning of the 20th century, wind erosion seriously damaged the soil in the Great Plains, parts of which became known as the 'Dust Bowl' during the 'Dirty Thirties'. The Friends of the Land, a group of scientists in fields such as soil protection, landscape development and ecology, promoted a sustainable way of farming that prevented erosion. Their journal *The Land* was intended to interest people in ecological and agricultural issues. Among the members of the group were two people who had an important influence on the early organic farming movement, Edward H. Faulkner and Louis Bromfield; the other members included the prominent ecologists Paul Sears and Aldo Leopold and the first head of the US Department of Agriculture's Soil Conservation Service, Hugh H. Bennett (Nelson, 1997).

The politician and part-time farmer Edward H. Faulkner (1886–1964) saw in *Plowman's Folly* (Faulkner, 1943) the roots of the erosion problem. He rejected the use of the mouldboard plough because the soil-turning effect destroys surface layers and structure of soils. Instead of ploughing he favoured a 'trash mulch system': he combined a surface layer of organic residues – so-called sheet composting – with ploughless soil cultivation to prevent erosion (Beeman, 1993a). In Pleasant Valley, Ohio, the novelist Louis Bromfield (1898–1956) experimented with sustainable farming during the 1940s (Bromfield, 1949; Beeman, 1993b). His Malabar Farm became a showpiece of organic farming. He linked organic farming with the romantic agrarian ideal of a 'Jeffersonian Republic': small, organic farms as 'cells' of a sustainable society.

Similar romantic agrarian ideas can be found in the urban American Food Reform movement during the 1940s and 1950s. Its activities – similar to those of Germany's Life Reform movement – concerned vegetarian food reform, back-to-the-land initiatives and organic gardening. A key figure in the movement was the editor Jerome I. Rodale (1898–1971), who started the magazine *Organic Gardening and Farming* in 1942. His book *Pay Dirt* was published in 1945 (Peters, 1979).

Curiously, although the writings of the prominent American soil biologist Selman A. Waksman were well known in Germany during organic farming's pioneer period (and were even translated into German), there are no hints that his scientific work on soil organisms (Waksman,

1930) and the humus economy (Waksman, 1926) played any role in the developing organic farming movement in the USA or the UK. Neither the writings of the organic pioneers in those countries nor the historical studies on the organic farming movement give much attention to his work.

***

When we examine organic farming of today in the light of the ideas and activities of the pioneers described here, we see that many of the important founding principles remain relevant. Yet organic farming has changed and developed over the decades. The remaining chapters portray these changes, depicting the evolution of the organic concept in response to the changing technological, political, economic and social environment.

References

Aubert, C. (1970) *L'Agriculture biologique*. Le courrier du livre, Paris.

Balfour, E.B. (1943) *The Living Soil*. Faber & Faber, London.

Baumgartner, J. (1992) *Ernährungsreform – Antwort auf Industrialisierung und Ernährungswandel – Ernährungsreform als Teil der Lebensreformbewegung am Beispiel der Siedlung und des Unternehmens Eden seit 1893*. Peter Lang, Frankfurt am Main, Germany.

Beeman, R.S. (1993a) 'The trash farmer': Edward Faulkner and the origins of sustainable agriculture in the United States, 1943–1953. *Journal of Sustainable Agriculture* 4, 91–102.

Beeman, R.S. (1993b) Louis Bromfield versus the 'Age of Irritation'. *Environmental History Review* 17, 77–92.

Bittermann, E. (1956) Die landwirtschaftliche Produktion in Deutschland 1800–1950. *Kühn-Archiv* 70, 1–145.

Bromfield, L. (1949) *Out of the Earth*. Harper & Brothers, New York.

Büsselberg, W. (1937) *Natürlicher Landbau – Bodenständige und gesunde Ernährung*. Deutsche Volksgesundheit, Nürnberg, Germany.

Conford, P. (1995) The alchemy of waste: the impact of Asian farming on the British organic movement. *Rural History* 6, 103–114.

Conford, P. (2001) *The Origins of the Organic Movement*. Floris Books, Edinburgh, UK.

Faulkner, E.H. (1943) *Plowmen's Folly*. Grosset & Dunlap, New York.

Glanz, F. (1922) *Die Wühlarbeit im Ackerboden*. Carl Gerold's, Vienna.

Görbing, J. (1947) *Die Grundlagen der Gare im praktischen Ackerbau*. Landbuch, Hannover, Germany.

Hensel, J. (1939) *Brot aus Steinen durch mineralische Düngung*, 3rd edn. Meyer, Leipzig. (1st edn. 1898).

Herr, F. (1927) *Bodenfruchtbarkeit und neuzeitliche Bodenbearbeitung*. Neulohe, Affoldern, Germany.

Hofstetter, M. (1942) *Neues Bauerntum – Altes Bauernwissen*. Selbstverlag, Ebmatingen, Switzerland.

Hopf, H. (1935) *Ackerfragen – Bodenleben und Ackergeräte*. Paul Parey, Berlin.

Howard, A. (1933) The waste products of agriculture: their utilization as humus. *Journal of the Royal Society of Arts* 82, 84–121.

Howard, A. (1935) The manufacture of humus by the Indore Process. *Journal of the Royal Society of Arts* 84, 25–59.

Howard, A. (1940) *An Agricultural Testament*. Oxford University Press, London.

Inhetveen, H. (1998) Women pioneers in farming: a gendered history of agricultural progress. *Sociologia Ruralis* 38, 265–284.

King, F.H. (1911) *Farmers of Forty Centuries*. Translated as *4000 Jahre Landbau in China, Korea und Japan* (1984). Siebeneicher, Munich, Germany.

Koepf, H.H. and von Plato, B. (2001) *Die biologisch-dynamische Wirtschaftsweise im 20. Jahrhundert*. Verlag am Goetheanum, Dornach, Switzerland.

Köhler, R. (1949) *Biotechnischer Ackerbau*. Bayerischer Landwirtschaftsverlag, Munich, Germany.

Kolisko, E. and Kolisko, L. (1939) *Agriculture of Tomorrow*. Kolisko Archive, Stroud, UK.

Könemann, E. (1925) Fiehloser Ackerbau – natürliche Bodenbearbeitung. *TAO* 11, 1–20.

Könemann, E. (1939) *Biologische Bodenkultur und Düngewirtschaft*, 2nd edn. Siebeneicher, Tutzing, Germany.

Krabbe, W.R. (1974) *Gesellschaftsveränderung durch Lebensreform*. Vandenhoeck & Ruprecht, Göttingen, Germany.

Krantz, H. (1922) Veredelung von Wirtschaftsdüngern – Bekohlungskraft – Saatguttüchtigkeit. *Deutsche Landwirtschaftliche Presse* 49 (87–88), 549–550; 49 (89–90), 561–562; 49 (91–92), 571–572.

Löhnis, F. (1910) *Handbuch der landwirtschaftlichen Bakteriologie*. Gebrüder Borntraeger, Berlin.

Martens, A. and Schwager, H. (1933) *Der neue Land-, Obst- und Gartenbau*. Lebensweiser-Verlag, Gettenbach, Germany.

McCarrison, R. (1936) Nutrition and national health. *Journal of the Royal Society of Arts* 84, 1047–1066, 1067–1083, 1087–1107.

McCarrison, R. and Viswanath, B. (1926) The effect of manural conditions on the nutritive and vitamin values of millet and wheat. *Indian Journal of Medical Research* 14, 351–378.

Moore-Colyer, R.J. (1999) Sir George Stapledon (1888–1960) and the landscape of Britain. *Environment and History* 5, 221–236.

Nelson, P.J. (1997) To hold the land: soil erosion, agricultural scientists, and the development of conservation tillage techniques. *Agricultural History* 71, 71–90.

Peters, S.M. (1979) The land in trust – a social history of the organic farming movement. PhD thesis, McGill University, Montreal, Quebec.

Pfeiffer, E. (1938) *Die Fruchtbarkeit der Erde, ihre Erhaltung und Erneuerung – Das biologisch-dynamische Prinzip in der Natur*. Zbinden & Hügin, Basel, Switzerland.

Remer, N. (1954) *Bodenständige Dauerfruchtbarkeit*. Forschungsring für biologisch-dynamische Wirtschaftsweise, Stuttgart, Germany.

Rudolph, W. (1925) *Der natürliche Landbau als Grundlage des natürlichen Lebens*. Fürs Land, Horben-Freiburg, Germany.

Rusch, H.-P. (1955) *Die Naturwissenschaft von morgen*. Hanns Georg Müller, Krailling bei München, Germany.

Rusch, H.-P. (1968) *Bodenfruchtbarkeit – Eine Studie biologischen Denkens*. K.F. Haug, Heidelberg, Germany.

Schaumann, W. (1994) 'Organismus' und 'Individualität' im Landwirtschaftlichen

Kurs Rudolf Steiners. *Lebendige Erde* 45, 315–325.

Schomerus, J. (1931) *Die Bodenbedeckung – Ein wertvolles Kulturverfahren.* C. Heinrich, Dresden, Germany.

Schuphan, W. (1937) Untersuchungen über wichtige Qualitätsfehler des Knollenselleries bei gleichzeitiger Berücksichtigung der Veränderung wertgebender Stoffgruppen durch die Düngung. *Bodenkunde und Pflanzenernährung* 2, 255–304.

Simons, G. (1911) *Bodendüngung, Pflanzenwachstum, Menschengesundheit – Ein Ratgeber für denkende Gartenfreunde,* 2nd edn. Lebenskunst-Heilkunst, Berlin.

Spiess, H. (1978) Konventionelle und biologisch-dynamische Verfahren zur Steigerung der Bodenfruchtbarkeit. PhD thesis, Justus-von-Liebig-Universität, Gießen, Germany.

Steiner, R. (1985) *Geisteswissenschaftliche Grundlagen zum Gedeihen der Landwirtschaft.* Rudolf Steiner Velag, Dornach, Switzerland.

Sykes, F. (1959) *Modern Humus Farming.* Faber & Faber, London.

Turner, N. (1951) *Fertility Farming.* Faber & Faber, London.

Vogt, G. (2000a) *Entstehung und Entwicklung des ökologischen Landbaus im deutschsprachigen Raum.* Ökologische Konzepte 99. Stiftung Ökologie & Landbau, Bad Dürkheim, Germany.

Vogt, G. (2000b) Ökologischer Landbau im III. Reich. *Zeitschrift für Agrargeschichte und Agrarsoziologie* 48, 161–180.

Wachsmuth, G. (ed.) (1929) *Gäa-Sophia – Band IV: Landwirtschaft.* Orient-Occident, Stuttgart, Germany.

Waksman, S.A. (1926) The origin and nature of the soil organic matter or soil 'humus'. *Soil Science* 22, 123–162, 323–333, 421–436.

Waksman, S.A. (1930) *Der gegenwärtige Stand der Bodenmikrobiologie und ihre Anwendung auf Bodenfruchtbarkeit und Pflanzenwachstum.* Urban & Schwarzenberg, Berlin.

3 Organic Values

M. Sligh[1] and T. Cierpka[2]

[1]*Director of Just Foods, RAFI-USA, PO Box 640, Pittsboro, North Carolina 27312, USA;* [2]*Director of Member Relations and Human Resources, International Federation of Organic Agriculture Movements, Charles-de-Gaulle-Str. 5, 53113 Bonn, Germany*

3.1 The Organic Alternative to Industrialized Agriculture

For much of the last century industrial agriculture reigned supreme. Indigenous and older forms of agriculture were viewed as hopelessly primitive and unworkable. Progress in food production meant more and larger machines, larger corporate farms and more chemical inputs. The fatal flaw of today's food production is that it is modelled on the industrial system. It does not attempt to remain within the bounds of nature but is rather designed to 'beat' nature: beat it with technology, cheap labour and externalization of costs.

The alternative is a food system that raises incomes and increases food security and food safety at both ends. It is one in which the environment is preserved, farmers and workers have fair access to the means of food production while receiving a fair return for their labour, and consumers have food they can trust at fair prices. These principles are the basis of organic agriculture, which sets out to be the fair, safe and sane alternative to the industrial model. But organic farming is at the crossroads and under enormous pressure to be like agribusiness. As we shall see, the very success of organic farming may be creating its greatest challenges ahead.

Organic agriculture is often described simply as a way of producing food and other products without synthetic fertilizers and pesticides. But as defined by the Codex Alimentarius Commission, the international food standards body established by the Food and Agriculture Organization of the United Nations and the World Health Organization, the concept has greater depth:

> Organic agriculture is a holistic production management system which promotes and enhances agro-ecosystem health, including biodiversity, biological cycles, and soil biological activity. It emphasizes the use of management practices in preference to the use of off-farm inputs, taking into account that regional conditions require locally adapted systems. This is accomplished by using, where possible, agronomic, biological and mechanical methods, as opposed to using synthetic materials, to fulfill any specific function within the system.
> (Codex Alimentarius Commission, 1999/2001)

This definition clearly asserts the environmental, locally appropriate and holistic nature of organic agriculture, but it does not fully reflect the full breadth of organic values. In contrast, the International Federation of Organic Agriculture Movements, the world's largest non-governmental organic organization, with members in every part of the globe, articulates a broader definition:

> Organic agriculture is an agricultural system that promotes environmentally, socially and economically sound production of food, fiber, timber, etc. In this system, soil fertility is seen as the key to successful production. Working with the natural properties of plants, animals and the landscape, organic farmers aim to optimize quality in all aspects of agriculture and the environment (IFOAM, 2003).

This definition places organic agriculture within the values of the broader organic community by requiring it to be socially just, economically viable and environmentally sound.

3.2 The Diverse Sources of Organic Values

The organic concept is commonly attributed to such European and American pioneers as Albert Howard and Eve Balfour in the UK, Jerome I. Rodale in the USA and Rudolf Steiner in Germany (see Chapter 2, this volume). Moreover, organic innovation has also been a farmer-led experience happening in different parts of the world simultaneously. This is a very important reason for its success. In particular, indigenous peoples have made important contributions to the organic approach, whether formally or informally; their survival and the preservation of their knowledge of techniques, practices and biodiversity will remain important for further organic development.

The organic approach is very ancient as well as modern and scientific. When Franklin H. King, chief of the US Department of Agriculture's Division of Soil Management, came back from his travels through China and other parts of East Asia in the early 20th century, he wrote admiringly about the permanent agriculture of the Far East in his book *Farmers of Forty Centuries* (King, 1911). King described an agriculture based on crop rotations, green manuring, intercropping, soil conservation and recycling of organic matter. The organic pioneers were very receptive to these ideas. Unfortunately, King died before his formal recommendations were realized. One can only imagine what modern agriculture might look like today if agricultural policy makers had heeded his experiences and advice!

Similarly, the European and American pioneers served as communicators to Western audiences of agricultural techniques that were not formalized but lived informally in farmer-based knowledge and wisdom. Howard confirmed this role: 'Howard was always wont to say that he learned more from the *ryot* [peasant] in his fields than he did from text books and the pundits of the classroom' (Watson, 1948). It is not a question of whether formal or informal knowledge is 'better'; they are different forms of knowledge, and both will continue to make valuable contributions to improving agriculture. Howard and other early Western organic visionaries played a critical role in connecting the formal with the informal, and the industrialized with the indigenous.

A lesser-known early organic advocate was Paul Keene, who in 1946 founded Walnut Acres, the oldest US mail order organic foods company. Keene went to India to teach English and came back a convert to organic farming, inspired by Mahatma Gandhi. In one of his books, *Fear Not to Sow*, he wrote an introduction entitled 'Inspiration from Gandhi', in which he stated his belief that organic farming was 'a priceless burden of trust ... that calls us to ever higher standards. To deal justly with the holy earth, with our foods, with the persons who work so hard to grow and prepare them and with the persons whose lives depend in part upon us' (Keene, 1988, pp. 2–4). Keene clearly frames organic agriculture within a broader value-based framework, but also within a systems approach, articulating a vision for organic agriculture that is committed to social justice, environmental protection, health and equity. He also was eloquent about not wanting to farm on such a large scale as to limit his ability to apply his idealism fully to all parts of his work.

3.3 The Breadth of Values in Organic Agriculture's Formative Years and Later

From the earliest days, as has already been discussed in Chapter 2, organic agriculture challenged the growing dominance of industrialized and corporate agriculture. This challenge is reflected, for example, in the writings of Jerome I. Rodale, founder of the magazine *Organic*

Gardening and Farming. Rodale wrote of the organic farmer as carrying a sacred trust, and encouraged farmers to become activists against bad governmental polices and giant vegetable factories. He laid out these broader organic values clearly in his book *The Organic Front* (Rodale, 1948). For him, organic agriculture started with the health of the soil and then spread across the whole food chain and into social values. He felt so strongly about these broader values of organic agriculture that he formed The Soil and Health Foundation, and called for 'legal action against one of these large vegetable factories, and get an injunction issued against such evil practices. The revolution has begun' (Rodale, 1948, p. 91). (No doubt he would be amazed to learn that many of those corporations have now bought US organic food companies in recent years to take advantage of the ever-growing consumer demand for healthier foods! [Sligh and Christman, 2003, p. 27].) Rodale also published Albert Howard's political book, *War in the Soil* (Howard, 1946), in which he spoke of the conflict between fresh, natural foods and agribusiness profits.

Pioneers such as Rodale are seen today by some as agricultural visionaries, but at first they were branded 'kooks', and organic farming remained very much on the fringe for a long time. The 'back to the land' movement of the 1960s and early 1970s embraced its ideas, and organic farming came to be perceived as 'hippie farming', and downright countercultural. This was a global phenomenon of primarily urban youth from 'Prague to Patagonia' who were becoming farmers and would-be farmers and were seeking out sane agricultural alternatives. Organic farming became associated with the cultural upheavals of this period and the public's questioning of the proper roles of governments and corporations. Young farmers wishing not to 'sell out' to the industrial–military complex chose organic farming, as did those who saw no role for government in agriculture or were concerned about the growing threats of corporate agribusiness. The farming techniques of the 'back to the land' movement of the 1960s and 1970s were based on organic principles, and the growth of organic foods through consumer food cooperatives came from these early organic farmers. Food cooperatives were based on the idea that food should be bought from as close by as possible, that the less processing and packaging the better and that consumers should organize themselves regarding food purchases to undermine the power of the food industry.

Some conventional farmers and farm workers also saw organic farming as a very practical solution to low commodity prices and wages, and as the best way to avoid toxic chemicals. Thus, there was a very broad set of values and hopes for organic agriculture that appealed to a wide range of stakeholders. Organic farming strove to be environmentally sound and locally rooted, a way for farmers to farm with dignity, and a way for family-size operations to be fairly compensated. It was agriculture 'with an attitude', whose basic principles combined values, techniques and, eventually, standards.

Acceptance of this broad conception of organic farming got considerable help around this time from related developments outside its immediate domain. A great blow to industrial agriculture and a big boost for the organic alternative came with the publication in 1962 of *Silent Spring*, by Rachel Carson (1962). Having worked as a marine biologist with the US Department of the Interior for 17 years, and having written several best-sellers on sea life, Carson became profoundly disturbed by the growing evidence of the environmental havoc caused by pesticides. The result was a book that sounded the alarm regarding the dangers of unchecked use of chemicals. *Silent Spring* set off a storm of international controversy. Although fighting terminal cancer, Carson took on her corporate opponents and did not back down. Her work launched a global movement against the misuse of pesticides, the very cornerstone of industrial agriculture, creating a new public awareness of the great dangers of the use of pesticides and helping to launch a consumer movement against the use of harmful chemicals in food. Its influence was not limited to pesticides; the book is widely regarded as a major force behind the more general environmental awakening of the late 1960s, in both the USA and many other countries.

A less direct example of how organic agriculture was affected by events going on in the larger world was the rise of the United Farm Workers in California in the late 1960s. Under the leadership of Cesar Chavez, the UFW organized tens of thousands of mostly Mexican migrant farm labourers into a union of solidarity and mutual support. The union's struggles garnered national attention, and its boycotts of California table grapes and lettuce became the focal point for mobilizing literally millions of consumers, as were its boycotts of agribusiness giants such as Dow Chemical and Del Monte (Sligh, 2002). A major demand of the union was an end to hazardous pesticide application practices that imperilled the lives of thousands of farm workers. Chavez's work also created a bridge between the interests of consumers and those producing and harvesting foods, a key relationship in the building of the organic movement.

The broad set of social, economic and environmental values underlying organic agriculture, sketched in Chapter 2, is clearly reflected in the history of the Soil Association in the UK (described in detail in Chapter 10, this volume), which was founded in 1946 and has provided vital continuity across the decades from the rise of organic farming to more recent organic activism. Its founders, who included farmers, agricultural scientists, nutritionists and many others, saw a direct connection between farming practices and the health of plants, animals, humans and the environment. Their catalyst was the publication in 1943 of Eve Balfour's *The Living Soil* (Balfour, 1943), which presented the case for a sustainable approach to agriculture as an alternative to the industrialized path that agriculture was increasingly following in the UK and elsewhere. Political and cultural considerations were important to the Soil Association from the beginning (Conford, 2001), and especially during the 1960s, its journal *Mother Earth* gave even greater attention to environmental as well as cultural issues (Chapter 10, this volume).

3.4 Organic Values and Integrity in the Face of Increasing Trade

Beginning in the early 1970s the organic movement evolved more fully from the pioneer phase with the formation of local organic farmer organizations in many countries. This happened in a very decentralized way at first, although the desirability of some international coordination was soon a driving force behind the founding in 1972 of the International Federation of Organic Agriculture Movements (IFOAM, described in detail in Chapter 9, this volume).

Among the roles of these organizations was to define consistent standards for organic food and establish programmes to certify farmers' compliance with them, as described in Chapter 8. They saw themselves as the defenders of organic integrity and values, and carefully nurtured loyal consumer confidence based on these values and principles.

However, as trade expanded, there was concern that working at the local level might not be enough to ensure the integrity of organic standards. Yet there also was a growing understanding that preserving integrity was critical for maintaining consumer confidence and for the continued growth of the organic movement. But organic foods no longer necessarily meant fresh and local, which had allowed organic integrity to be based on consumers' confidence in local producers; organic foods were also entering national and global trade channels (as discussed in Chapter 7, this volume). This meant that greater consistency was needed among the various local standard-setting bodies, and led to one of the most challenging and critical periods of the organic movement, that of national and international institutionalization and harmonization of standards.

For example, the European Union put organic regulations into effect in 1992; in 2001 the United Nations' Codex Alimentarius Commission adopted global organic guidelines; in the USA, national standards were implemented in 2002 that replaced the (somewhat) diverse standards of dozens of separate certifying bodies (see Chapter 8, this volume, for a more detailed account of these developments). The purpose in all cases was to ensure continued growth in organic trade and to maintain consumers' confidence in the organic label.

This development has put a great strain on the underlying values and principles of organic farming, which initially were embedded in a much more locally oriented vision of agriculture. In response, in 2004 IFOAM started a unique consultation among global stakeholders to define the principles of organic agriculture (Luttikhult, 2007). This process, which lasted for 18 months and ended with the approval of IFOAM's General Assembly in 2005, stated the basis of organic agriculture as follows (IFOAM, 2005).
Organic agriculture is based on the:

- Principle of health;
- Principle of ecology;
- Principle of fairness;
- Principle of care.

These principles are the roots from which organic agriculture will continue to grow and develop. They reflect the contribution that organic agriculture can make to the world and a vision to improve all agriculture in a global context. (Chapter 8, this volume, gives the explanations accompanying these principles, as well as statements preceding earlier IFOAM standards comprising 7 and 15 principles, respectively.)

Clearly, organic food is recognized as a significant alternative to industrialized agriculture. However, it remains to be seen whether the integrity of the institutionalized organic standards can be maintained (a question discussed in Chapter 8, this volume, as well). As the critical debate continues regarding what is organic, it is vital for the key underlying values that embody the integrity of the organic alternative to be (re)affirmed, and then aggressively protected and vigorously maintained.

3.5 Key Elements of Organic Integrity

Organic agriculture is associated with sound environmental stewardship and with social justice and improvements in the quality of life for those participating. It is associated with fairness, transparency and 'doing the right thing'. It is associated with values that include improved health, food safety and quality, and worker safety. Finally, it is associated with transforming agriculture from its present system to one in which people live within the bounds of nature. These associations help reveal the following key elements of organic integrity.

3.5.1 Environmental stewardship

This includes production and processing systems that promote and enhance biodiversity and ecological balance. It starts with the health of the soil and embraces a whole systems approach. Organic integrity has endured because it does not support expedient short cuts that in the long run damage the environment. Environmental values were clearly seen very early on as important elements of the organic approach. Appropriate stewardship raises an important challenge for organic farming: to uphold fundamental, universal organic principles while adapting production systems appropriately for different environments.

Organic agriculture has not yet fully confronted the environmental dangers posed by genetic engineering and the patenting of biological processes, although there is increasing activity to develop and preserve germplasm for organic agriculture as the non-GMO alternative. Only recently has it started to address the big questions of sustainable energy use. A study issued by the IFOAM (2004) about organic agriculture and climate change presents many benefits of organic farming compared with those of conventional agriculture that have already been explored. However, the study also underlines the need to invest in serious crop- and climate-specific research. Such research could show organic agriculture's potential contribution to countering global climate change by increasing carbon sequestration. Furthermore, energy use in the entire supply chain, including production, transportation and processing, needs to be analysed in more detail. Such an analysis could strengthen the organic community's historical support for local production and consumption cycles.

3.5.2 Accountability and fairness

The lifeblood of organic agriculture is grass-roots, consumer-based confidence in, and demand for, safe foods that are produced and processed using environmentally sound, humane and socially just practices. These must be based on public transparency, honesty and direct consumer access. Organic integrity also requires accountability to local communities for the impacts of our organic production and processing on local, regional and international economies. Organic integrity embraces the promotion of fair trade practices, which support local food systems, family farms and food security. It cannot allow organic colonialism or any other practices that perpetuate historically unjust relationships between nations of the global North and those of the South. Processors and retailers will have to see that it is in their enlightened self-interest to support these principles and take a greater share of the risks and costs associated with the organic approach. Good examples in this context are Lebensbaum and Rapunzel in Germany, Stassen in Sri Lanka and Sekem in Egypt, where accountability and fair pricing systems are part of each company's profile.

When organic standards became significant in the late 1970s through the mid-1980s, little attention was given to social justice, no doubt because there did not seem to be any pressing need to do so. At the time, sales of organic products from the global South to the North were not significant, and hence the question of fair trade was not relevant. Corporate megafarms had not made serious inroads into organic farming, so the industrialization of agriculture – a major concern of many organic pioneers – did not seem to pose any great threat within the organic sector. Because organic farms were typically small, they hired few workers, and so farm labour standards were not an urgent concern.

However, all that changed as organic farming grew in importance worldwide. Thus, in 1992, the IFOAM General Assembly in São Paulo decided to develop a chapter on social justice for IFOAM's Basic Standards. It took 4 years until the 1996 General Assembly in Copenhagen approved such a chapter as draft standards. In 2002, implementation of the social justice chapter became binding for all IFOAM Accredited Certifiers. The process took so much time because the movement was split over the issue: grass-roots members from the South considered it very important to define the social dimension of organic agriculture in detail, whereas the certifiers – mainly from the North – argued against detailed standards because of increased costs for inspection and certification and the problem of 'inspectability' of social standards. However, many founding members of the fair trade movement have long been part of the organic movement, and finally convinced the organic community that social justice is an integral part of organic production and processing. Simultaneously, US domestic groups were developing a marketplace approach for recognizing farmers and their workers who were meeting the social values of organic farming (Henderson et al., 2007).

The early organic visionaries saw organic farming as an alternative to agribusiness as usual. They set out to combine environmental stewardship, accountability and fairness into an alternative model, as well as to set enlightened labour standards for other food systems to strive for. This movement started out not to produce expensive niche-market foods for rich people, but rather to offer a model for all of agriculture. Expanding this accessibility for all peoples should be a top priority. Fair prices and the rights of farmers and farm workers are essential for this goal to be achieved. The real promise of organic agriculture is to provide credible production systems that can aid the world's poor and strengthen local food security.

References

Balfour, E.B. (1943) *The Living Soil*. Faber & Faber, London.

Carson, R. (1962) *Silent Spring*. Houghton Mifflin, Boston, Massachusetts.

Codex Alimentarius Commission (1999/2001) *Guidelines for the Production, Processing, Labelling and Marketing of Organically Produced Foods*. CAC/GL 32-1999/Rev 1, 2001, Rome.

Conford, P. (2001) *The Origins of the Organic Movement*. Floris Books, Edinburgh, UK.

Henderson, E., Mandelbaum, R., Mendieta, O. and Sligh, M. (2007) *The Agricultural Justice Project: Social Stewardship Standards in Organic and Sustainable Agriculture*. Rural Advancement Fund International – USA, Pittsboro, North Carolina.

Howard, A. (1946) *War in the Soil*. Rodale Press, Emmaus, Pennsylvania.

IFOAM (2003) *IFOAM Annual Report 2002*. International Federation of Organic Agriculture Movements, Tholey-Theley, Germany. Available at: www.ifoam.org/about_ifoam/inside_ifoam/pdfs/IFOAM_Annual_Report_2002.pdf

IFOAM (2004) *The Role of Organic Agriculture in Mitigating Climate Change*. International Federation of Organic Agriculture Movements, Bonn, Germany.

IFOAM (2005) The principles of organic agriculture. International Federation of Organic Agriculture Movements, Bonn, Germany. Available at: www.ifoam.org/about_ifoam/principles/index.html

Keene, P. (1988) *Fear Not to Sow*. The Globe Pequot Press, Chester, Connecticut.

King, F.H. (1911) *Farmers of Forty Centuries*. Reprinted 1973 by Rodale Press, Emmaus, Pennsylvania.

Luttikholt, L.W.M. (2007) Principles of organic agriculture as formulated by the International Federation of Organic Agriculture Movements. *NJAS* 54, 347–360.

Rodale, J.I. (1948) *The Organic Front*. Rodale Press, Emmaus, Pennsylvania.

Sligh, M. (2002) Organics at the crossroads: the past and the future of the organic movement. In: Kimbrell, A. (ed.) *The Fatal Harvest Reader*. Island Press, Washington, DC, pp. 272–282.

Sligh, M. and Christman, C. (2003) *Who Owns Organic? The Global Status, Prospects, and Challenges of a Changing Organic Market*. Rural Advancement Foundation International – USA, Pittsboro, North Carolina. Available at: www.rafiusa.org/pubs/OrganicReport.pdf

Watson, E.F. (1948) The lessons of the East. *Organic Gardening Magazine*, 13 September. Available at: journeytoforever.org/farm_library/howard_memorial.html#Watson

4 The Science of Organic Farming

D.H. Stinner

Research Scientist and Administrative Coordinator, Organic Food and Farming Education and Research Program, The Ohio State University, Ohio Agriculture Research and Development Center, 1680 Madison Ave., Wooster, Ohio, 44691, USA

4.1 The Relationship Between Organic and Mainstream Agricultural Research

Organic farming's relationship to science is deeply imbedded in its philosophical and cultural roots as a reaction against the industrialization of agriculture in the early and mid-20th century. The agricultural scientific establishment, using the powerful tools of reductionistic experimentation in the context of the ideas of Justus von Liebig (1842), led to the development of a new materialistic and mechanistic agriculture based on chemistry. The intellectual principles and values underlying this agricultural revolution contrasted sharply with those of organic farming discussed in Chapter 2, e.g. the farm as a self-regulating organism that should mimic nature and function in harmony with its environment (Steiner, 1974), and the importance of healthy and vital soils as the foundation of healthy crops, animals and people (Howard, 1947; Balfour, 1948, 1978; Rodale, 1948; Steiner, 1974; Voisin, 1999). As a result, throughout its history and to a lesser extent continuing to this day, some advocates of organic farming to varying degrees have felt distrust or even disdain for the agricultural scientific establishment and its Baconian/Cartesian scientific methods, which they considered inadequate to address the critical issues in agriculture (e.g. Balfour, 1948; Steiner, 1974; Baars, 2002). In recent years, however, a good deal of organic research has come to resemble conventionally oriented research much more closely (Lockeretz, 2000).

Despite the sharp initial separation of organic farming from mainstream agricultural science, the organic movement's development has by no means been unscientific, as seen in Chapter 2. Indeed, many scientists in various disciplines contributed significantly to the emergence of organic farming and were among its leaders from its beginnings (Woodward, 2002). However, the research methods advocated by some of the early scientific leaders contrasted strongly with the methods commonly accepted at the time and today.

This contrast can be seen, for example, in the thinking of three especially influential pioneers: Rudolf Steiner, Albert Howard and Eve Balfour (all of whom have been discussed in Chapter 2, this volume). Steiner's suggestions for research on biodynamic agriculture were strongly influenced by Goethean science and phenomenology (Steiner, 1974; Baars, 2002; van Steensel et al., 2002). These ideas led to the development of picture-forming methods to distinguish organically and conventionally produced foods (Pfeiffer, 1936, 1984; Kolisko and Kolisko, 1939); these methods are not generally accepted in orthodox agricultural circles today. Furthermore, in contrast to the conventional experiment station model of agricultural scientists conducting short-term disciplinary trials on one or at most a few management factors, Steiner suggested direct linking of the scientific knowledge of disciplinary teams of scientists with the empirical knowledge of farmers, and placing scientific knowledge in the context of working farms and long-term whole-farm studies (van Steensel et al., 2002). However, unlike picture-forming methods, whole-farm research with farmers' participation has gained in popularity in mainstream circles in recent years.

Related views offered by Howard became a hallmark of the science of organic farming, namely the distinction between specialized and holistic research:

> In considering ... the large volumes of scientific papers dealing with manurial questions, which have been poured out of experiment stations for the last fifty years, we have been impressed by the evils inseparable from the present fragmentation of any large agriculture problem. ... All this seems to follow from the excessive specialization which is now taking place, both in teaching and the application of science. In the training given to the students and in much of the published work, the tendency of knowing more and more about less and less is every year becoming more marked. For this reason, any review of the problem of increasing soil fertility is rendered particularly difficult not only by the vast mass of published paper but also by the fragmentary and piecemeal nature.
>
> (Howard and Wad, 1931, p. 9)

Howard offered specific suggestions for how agricultural research should be conducted:

> We must emancipate ourselves from the conventional approach to agricultural problems by means of the separate sciences and above all

from the statistical consideration of the evidence afforded by the ordinary field experiment.

(Howard, 1943, p. 22)

He suggested three principles on which agricultural research should be based. First, it should take 'a synthetic approach and look at the wheel of life as one great subject and not as if it were a patchwork of unrelated things' (p. 22). Second, 'the researcher should be deeply trained in all the sciences, with the training to include travel to understand a diversity of conditions and to acquire intimate practical farming knowledge and a deep respect for local farmers and their knowledge' (p. 221). Third, 'the proof of the research should be in a working farm context, which cannot be simulated in small plots' (pp. 185–186).

Similarly, Balfour (1948) and her scientific advisers, out of frustration with the scientific agricultural establishment of their time, laid out a framework for long-term, whole-farm and interdisciplinary organic farming research that she felt was needed to scientifically test her thesis that healthy soils are the foundation of a healthy food chain of plants, animals and people. This research framework was manifested in the famous 'Haughley experiment' (Balfour, 1948, 1978), which contrasted sharply with replicated small-plot experiment station research of that time (e.g. at Rothamsted) in its holistic/farming systems perspective.

The criticisms offered by Steiner, Howard, Balfour and others have pervaded at least the rhetoric of organic research ever since, although whether they are reflected in how most organic research is actually done these days is debatable (Lockeretz, 2000). From the other side, it is not surprising that mainstream agriculture historically has felt a similar distrust and disdain for the scientific ideas underlying organic farming (e.g. Jukes, 1981a,b; DeGregori, 2004). This tension between conventional and organic concepts can be seen in the scientific and popular literature of the movement throughout its history, and has moulded the character of organic farming research worldwide (e.g. Niggli, 1999; Baars, 2002; Delate, 2002; van Steensel et al., 2002). However, as will be clear in the rest of this chapter, the gap between the approaches taken in organic and conventionally oriented research has substantially narrowed – too much, some might say. However, to enable this development to be better appreciated, I will first discuss in more detail the views of these three important pioneers.

4.2 Early Ideas About Organic Farming Research

4.2.1 Steiner and biodynamic agriculture

At the request of farmers, veterinarians and others involved in the anthroposophical movement who observed alarming declines in both crop and livestock quality and performance (see Chapter 2, this volume), Steiner delivered a series of eight lectures on agriculture in 1924 (Steiner, 1974) that became the foundation of the worldwide movement of biodynamic agriculture. He intended these lectures to provide practical solutions to problems, and not theoretical constructs. His solutions draw on the spiritual scientific perspective he developed (Steiner, 1971) and involve ideas far beyond what most agricultural scientists would consider scientific in any sense. However, these ideas and biodynamic agriculture have come to play an important role in the history of organic farming and associated scientific studies, as his followers and others have used the scientific method to explore and evaluate biodynamic methods.

Steiner emphasized the importance of unseen cosmic forces on crops, animals and soil, and the importance of incorporating these forces in properly manuring the soil to restore the vitality necessary for production of healthy crops and livestock. His prescriptions pre-date later verification of powerful effects that very minute quantities of certain substances can have on plant growth, and are similar in some respects to the homeopathic tradition in human and animal medicine. He stressed the importance of organic matter, humus and carbon (C), as well as nitrogen (N), organic sources of potassium, calcium, sulphur, silicon and phosphorus (P), and soil fauna, in particular earthworms. He conceptualized farms ideally as self-sufficient, but interacting with the whole of life, the earth and the cosmos.

Steiner stressed the relationships among different landscape elements on farms and the surrounding environment for healthy functioning. As such, his ideas were extremely holistic, and in many ways set the tone for what would become a hallmark of organic farming that sets it apart from the conventional scientific agriculture stream:

> What does science do nowadays? It takes a little plate and lays a preparation on it, carefully separates it off and peers into it, shutting off on every side whatever might be working into it. . . . It is the very opposite of what we should do to gain a relationship to the wide spaces. . . . We must find our way into the macrocosm. Then we shall once more begin to understand Nature – and other things too.
> (Steiner, 1974, p. 119)

Building on Steiner's concepts, Ehrenfried Pfeiffer and co-authors published several editions of *Bio-dynamic Farming and Gardening: Soil Fertility Renewal and Preservation* (Pfeiffer, 1938, 1940, 1943). The 1938 edition describes the then growing dominance of 'scientific agriculture', with its emphasis on chemical/mineral fertilization, mechanization and overall industrialization, in contrast to an approach founded on 'the principle of an *Organic Whole*', in which the soil, field and farm are viewed as living organisms; and in practice on humus production and maintenance through 'good manuring as . . . the basis of all agriculture' (Pfeiffer, 1938, p. 23).

We see in these works considerable scientific development of the argument that simply adding macronutrients in mineral form to increase production and to replace those taken by previous crops is not sufficient to sustain the capacity of soils to produce healthy crops, livestock and humans. Pfeiffer (1938, chs 4 and 5) stresses the importance of soil biological processes in the formation and maintenance of humus and soil function in general, especially with respect to earthworms and soil microorganisms and their roles in organic matter decomposition, nutrient release and retrieval from deep stores in the soil profile. Indeed, we see in his 1938 book a deep qualitative and sometimes quantitative understanding of what much future organic farming research would focus on – soil ecology. He offered a blend of informed discussion and presentation of actual research results (albeit without statistical measures of treatment effects) from many experiments designed to compare how various biodynamic methods with mineral fertilization, manure without biodynamic treatment and controls affected many different characteristics of numerous crop and animal species. He thereby provided considerable support for manure-based fertility supplementation in general as opposed to mineral fertilizers, and for Steiner's (1974) supposition that crops and animals can benefit from very subtle environmental influences involved in biodynamic compost and preparations (see Chapter 2, this volume, for a discussion of the ambiguous interpretation of results of such studies).

Of particular interest are the animal health studies with biodynamically versus minerally fertilized feeds that show striking differences in feed preferences, egg production, and disease resistance and resilience in young turkeys, with the biodynamic treatments being better overall. Pfeiffer (1938, p. 188) stressed the importance of proper experimental preparation for this type of research, with particular emphasis on establishing the land in the contrasting management methods for sufficient time (several years) to give the soil time to 'develop', and establishing common baseline conditions in experimental animals before implementing a specific experiment and then very carefully ensuring that all variables except the fertilizer treatment, for example, are the same for all treatments.

Another interesting idea that Pfeiffer (1938, p. 201) raised was a negative feedback loop within a farm, with poor-quality humus in the soil resulting in poor-quality livestock feed, which in turn results in poor-quality manure to return to the soil. Finally, in addition to providing considerable scientific data and information from component-oriented, replicated and often repeated experiments, he also provided data from several farms across a range of soil and climatic conditions in Central Europe before and for some years after conversion to biodynamic management. Overall, these data showed that biodynamic methods sustained and even enhanced soil fertility and usually increased yields (Pfeiffer, 1938, ch. 11).

4.2.2 Albert Howard and soil organic matter

Howard, considered by many in the English-speaking world to be the father of organic farming, wrote numerous books and scientific articles. In 1931, he published his most important scientific publication, *The Waste Products of Agriculture: Their Utilization as Humus* (Howard and Wad, 1931). This book was based on 26 years of studying improved crop production in Indian smallholdings by efficient use of wastes for improving and maintaining soil fertility. Howard was greatly influenced by Asian peasant farming traditions (Conford, 1995) and combined scientific principles with indigenous knowledge and experience. This book is very important in the history of the science of organic farming because in it Howard lays a scientific foundation, based on published research of the time, of the nature and role that soil organic matter (SOM) and, in particular, soil humus play in soil fertility. He explains the then recent understanding of the importance of soil microorganisms in the various steps during the formation of soil humus, from the decomposition of fresh plant and animal remains and their organic constituents (sugars, starches, pectins, celluloses, proteins, amino acids, lignins, etc.) to the production of available N for crop uptake by the slow oxidation of humus. Also included are reviews of research on losses of N during organic matter decomposition and on various sources of organic matter and their decomposition properties. All this basic information was used to lay a scientific foundation for the method of Indore composting (Howard and Wad, 1931).

In 1940, Howard published his famous book, *An Agricultural Testament*, which was his first book aimed at the general public. In it he explains a fundamental principle of organic farming: that humans should farm and manage soils after 'Nature's methods of soil management', with careful attention to balancing processes of growth and decay by incorporation of livestock. From this foundation, he provides evidence of the importance of humus in soil aggregation and aeration and discusses how the loss of humus through chemical farming influences soil erosion, diseases and pests of crops, livestock and humans, and the production, taste, quality and keeping properties of agricultural products, with examples based on his years of research in India, particularly with respect to his Indore process of composting. He brings to the scientific understanding of organic farming the important role of mycorrhizal fungi in the nutrition of many plants and the dependence on, and interaction of, these fungi with humus. He also points out the importance of the N cycle and the need to understand its local function, particularly with respect to periods of peak nitrate production and leaching potential for the development of the most optimum crop production management strategies and crop rotations.

In 1945, Howard published his *Farming and Gardening for Health or Disease* (Howard, 1945), also for a general audience. The book was republished in 1947 in the USA under the title *The Soil and Health: a Study of Organic Agriculture*. In this book, Howard chronicles his own research history and the development of his thinking on topics from agricultural investigations to public health systems. He reiterates the need for agricultural scientists and farmers to look to the operation of nature and argues that nature's 'great Law of Return' has been ignored by the various schools of agricultural science of the day. He presents an overview of the dominant forms of agriculture used at the time from both the East and the West, and a history of agriculture in Great Britain.

Howard attacks reductionistic science and the growing trends of specialization and separation of theory from practice. The famous Broadbalk wheat trials at Rothamsted, which were used to support replacement of organic sources of fertility with chemical fertilizers in much of the Western world's agriculture during this time, receive particular criticism. Diseases of the soil and of many of the world's dominant crops (as well as some of their insect pests) are reviewed, as are disease and health in livestock and humans. In all cases he suggests that 'much of this disease is due to farming and gardening methods which are inadmissible' (Howard, 1947, p. 187).

He argues that agriculture needs to replicate the forest to support healthy crops, livestock and humans sustainably (see Barton, 2001, for a discussion of the historical relationship that Howard's ideas as presented in his 1945 book and the roots of the organic movement have to early forest conservation). Freshly prepared humus and all its derivatives are the only substitute for the forest for humans living outside hunting–gathering cultures, Howard argues. He ends this book with the argument that this is the foundation of a healthy and vital civilization:

> One of the great tasks before the world has been outlined in this book. It is to found our civilization on a fresh basis – on the full utilization of the earth's green carpet. This will provide the food we need: it will prevent much present-day disease at the source and at the same time confer robust health and contentment on the population.
>
> (Howard, 1947, p. 261)

4.2.3 Balfour and the Haughley experiment

The Living Soil (1948) by Balfour is one the most important foundational books in English for the science of organic farming. Writing for both a lay and professional audience, Balfour draws on a wide spectrum of existing direct scientific evidence and indirect evidence for her contention that the basis of human health is management of soil humus in such a way that nature's 'law of return' is followed. More specifically, as part of her argument she brought considerable scientific knowledge to support the hypothesis shared by her and Howard that mycorrhizal fungi in humus play a key role in the chain of healthy soil–healthy plants–healthy livestock–healthy humans. She also reviewed nutrition research conducted by Robert McCarrison, a doctor, early forest soil ecology research by botanist Rayner and Howard's agricultural research.

Of particular historical importance is that we see in these ideas the foundation of what would become a strong scientific link between soil ecology and organic farming. Additional support for her argument was offered in the form of indirect evidence based on widespread practical knowledge and information on the diet and soil management of several widely varying indigenous peoples well known for their outstanding health (e.g. the Hunzas of north-western India). From this she concluded that what confers such health is a combination of consumption of whole foods (rather than processed foods) and the incorporation of some form of compost in their agricultural production systems. Balfour summarized her view of health under five propositions:

1. The primary factor in health (or the lack of it) is nutrition.
2. Fresh unprocessed natural whole foods (such as whole wheat bread, and raw vegetables and salads) have a greater nutritive value than the same foods when stale, or from which vital parts have been removed by processing or have been destroyed by faulty preparation.
3. Fresh foods are more health promoting than preserved foods (dried, canned or bottled).
4. The nutritive value of food is vitally affected by the way in which it is grown.
5. An essential link in the nutrition cycle is provided by the activities of soil fungi, and for this and other reasons the biological aspects of soil fertility are more important than the chemical.

(Balfour, 1948, ch. VII)

These remain basic tenets of the organic movement. To provide further scientific evidence of their validity, Balfour and other supporters organized the Haughley experiment in 1939 (although its actual start was delayed by World War II). This was the first large-scale organic 'systems' experiment to compare organic farming and conventional chemical-based farming (its establishment and administration are described in Chapter 10, this volume). This experiment reflected classic organic thinking about how scientific research should be conducted, a way of thinking that permeates organic farming research to this day. It also demonstrated the challenges of true systems research that still create problems for organic researchers. In contrast to the dominant small-plot research framework at Rothamsted and other government-supported agricultural research stations, this privately funded project was designed to create a large-scale holistic experimental framework in which the long-term effects of humus on health could be demonstrated conclusively to scientists and governments. To this end, the Haughley Research Trust was founded in 1938 by Alice Debenham 'to investigate the causes of positive health in crops and livestock, and particularly the relationship between the health of the soil and that of the crops and animals raised upon it' (Balfour, 1948, ch. VIII).

In 1940, 85 ha of land, partly owned by the Trust and partly leased, was divided into sections representing three farming systems. One was the Organic Section, where only animal manure from the system's livestock and vegetable residues in the form of organically prepared compost were used. Second was the Chemical Section, in which inorganic chemical fertilizers were supplemented by ploughing-in green crops (no animal manure was used). Finally, the Mixed Section, like the Organic Section, had crops and livestock, but farmyard manure and compost were used in conjunction with both organic and inorganic fertilizers. The Organic and Mixed Sections had the same classes of livestock of common origin.

According to Balfour (1948, ch. VIII):

> [T]he distinguishing, and in some respects novel, feature of the Haughley method of research will be the comparison of contrasted systems of soil management
>
> 1. Over a period of years, so that continuity is secured,
> 2. Over successive generations of plants and animals nurtured in the same way, so that cumulative effects may have full play,
> 3. On a regular rotational basis and on a field scale, so that all other conditions may be those of ordinary farming practice, and
> 4. On areas of land as nearly as possible comparable as regards soil type, drainage and other basic factors.

Further issues that set a precedent for future organic farming research included experimental design with systems of differing rotations. The experiment's managers concluded that:

the attempt to have identical rotations and to make a year by year, field by field, comparison is the wrong policy for the following reasons:
(a) Basically the experiment was designed to compare three methods of farming. Each section, therefore, should represent, as far as possible, the very best farm of its type. Clearly a rotation which was right for one method of farming would not necessarily be right for another; (b) In practice, it would not be possible to carry out all field operations simultaneously in all three sections, so that an accurate field by field comparison would in any case seldom be possible.
(Balfour, 1948, ch. VIII)

Furthermore, in view of the above issues it was decided that the only comparisons, including economic comparisons, would be over the entire rotation period. Interestingly, in spite of general opposition to the philosophy and research approach at Rothamsted, the Soil Association booklet contains a statement about being 'indebted to the National Research Station at Rothamsted for valuable co-operation and advice' (Balfour, 1948, ch. VIII).

It was decided that if soil fertility and cropping experiments were to be carried to their logical conclusions, feeding experiments on farm animals as well as laboratory animals would be needed. 'These will have to be continued through many generations of crops and livestock. So far as can be discovered, there is no provision at existing research stations for this integration of manuring and feeding through successive generations. Yet without it, no experiment seeking to determine ultimate food values can be complete' (Balfour, 1948, ch. VIII).

Here we see the foreshadowing of contemporary scientific issues related to the effects of organic management on food quality. When the Soil Association took over the Haughley Research Farms in 1948, it was decided that the experiment should be designed to compare the three methods of farming with regard to three consequences: quality of the resulting food; resistance of plants and animals to disease; and fertility of seed and animals (Balfour, 1948, ch. VIII).

Balfour (1948, ch. VIII) provided some results available early in the experiment. She reported that by the sixth generation of wheat, the proportion of small withered grain in the chemical system was much greater than in the organic system. Longer root lengths in wheat and beans were observed in the organic system than in the chemical or mixed ones. Barley yielded more in the chemical system, probably because of its fast growth rate and ready uptake of easily available mineral nutrients in the chemical fertilizer. In a potato experiment with two varieties it was noted that compost produced a higher proportion of large potatoes. Finally, in a comparison of different farmyard manures and composts, well-made compost produced Brussels sprouts with the best size and colour and the least insect attack. However, in both the potato and Brussels sprout comparisons, other responses to the treatments were mixed.

Looking back on the entire experiment in 1977, in a speech to the first scientific conference of the International Federation of Organic Agriculture Movements in Switzerland, Balfour said that among the most important and surprising findings was that levels of available soil nutrients fluctuated seasonally, such that maximum levels correlated with maximum plant demand in all three farming systems studied. However, up to 10 times more available P was found during the growing period than the dormant period in the Organic Section (in a field particularly high in organic matter), a much greater fluctuation than in the Mixed or Chemical Sections. Results for N and K were similar (Balfour, 1978, p. 20).

The absence of consistent differences in chemical analyses of crops or livestock products, except for usually higher water content of the chemically grown fodder, was interpreted as an indicator of the sustainability of the organic system, which was a closed-cycle system. This prompted Reginald Milton, the biochemist working on the project, to state that 'the analytical work carried out in connection with the Haughley Experiment has shown how wasteful of natural resources is modern commercial farming and how with a closed-cycle technique nutrients are recycled and moreover become available in situ – provided that an ecological approach is made to the methods of cultivation and farm management' (Balfour, 1978, p. 20). Balfour suggested that the difference in root system distribution, with greater lateral root branching at the expense of deep rooting exploration under chemical fertilization compared to organic fertilization, is another important functional difference between the systems.

By modern scientific standards, the Haughley experiment was more of a demonstration than a true experiment because there were no replicates. Therefore, it is not possible to evaluate the statistical significance of differences among the systems. The Haughley experiment ran into financial problems in the 1950s. Herein lies the challenge that would face future organic farming systems research: to find the resources to construct and sustain replicated whole-farm systems in an experimental context with sufficient management and scientific expertise to produce truly meaningful data.

4.3 The Rise of Organic Research Institutions

In the earliest days of organic farming, there were few research facilities, mainly private (e.g. the Haughley experiment), with virtually no public support for research. That situation largely remained unchanged until the 1970s, when an organic research infrastructure began to be created, slowly at first, but much more rapidly in recent years. A small sample of significant developments has been chosen to reflect the variety of missions and institutional arrangements that characterized early organic research activities.

4.3.1 Biodynamic research

Biodynamic research has the longest history, with its beginnings in the 1920s. Research centres include: the Natural Sciences Section of the Goetheanum in Dornach, Switzerland; the Research Institute of Biodynamic Agriculture in Darmstadt, Germany (1954); the Michael Fields Institute in East Troy, Wisconsin, USA (1984); and the Biodynamic Research Institute in Järna, Sweden (1986). Chapter 2 describes the principles that characterize biodynamic agriculture in contrast with other forms of organic farming.

4.3.2 The Rodale Institute, Pennsylvania, USA, 1947

The Rodale Institute (originally the Soil and Health Foundation) was the first organization in the USA concerned with advancing organic farming knowledge. It was founded by J.I. Rodale shortly after he became familiar with the ideas of Balfour, Howard and other British organic pioneers (see Chapter 2, this volume). A guiding theme in its work was the fundamental organic principle that healthy soil in turn produces healthy plants, animals and people.

Research efforts expanded considerably in the 1970s with the acquisition of a 135 ha research farm in Kutztown, Pennsylvania. A particularly significant research activity at that site has been the Farming Systems Trial, established in 1981 to compare a 5-year organic manure-based rotation, a 4-year organic cover crop rotation and a conventional 2-year cash grain rotation.

Although originally concerned specifically with organic methods, in later years the Rodale Institute dealt with various related but different alternatives, namely 'low-input', 'sustainable' and 'regenerative' agriculture.

4.3.3 The Research Institute of Organic Agriculture (FiBL), Frick, Switzerland, 1973

FiBL (available at: www.fibl.org) was established by organic farmers and scientists as a private foundation in 1973. Its mission was to conduct research projects and consultancy to support organic farmers, who at that time were not being served by the federal and canton authorities. Its many research programmes cover fruit, viticulture, arable farming, dairying, livestock and bees on its farm in Frick and on more than 200 working farms throughout Switzerland. Particularly noteworthy is its DOK trial, started in 1978, which compares biodynamic, organic and conventional systems (the history and activities of FiBL are described in more detail in Chapter 14, this volume).

4.3.4 Louis Bolk Institute, Driebergen, the Netherlands, 1976

The Louis Bolk Institute (available at: www.louisbolk.org) was founded in 1976 as a non-profit foundation to link social issues with research on organic and sustainable agriculture, nutrition and health care. It employs a broad range of researchers, from soil scientists to physicians, who work intensively with other research institutes at home and abroad. Research projects draw on the practical and experiential knowledge of hands-on professionals such as farmers, doctors and therapists, and place this knowledge in a wider context and provide it with a scientific basis. The Institute describes its goals as helping farmers with practical solutions for farm management, providing greater insight into healthy nutrition, helping doctors to promote human health and vitality, and helping researchers throughout the world with scientific innovation.

4.3.5 First IFOAM Scientific Conference, 1977

At its General Assembly in Seengen, Switzerland, in 1976, the International Federation of Organic Agriculture Movements (available at: www.ifoam.org), which had been organized in 1972 (as described in detail in Chapter 9, this volume), decided to hold a scientific conference in autumn 1977. The aim of the conference – the first of its kind in the world – was to give an overall view of the situation of research on organic agriculture. It was considered important to include reports of research at the planning stage as well as ongoing projects, with the programme to be kept as broad as possible to enable participants to coordinate their research with other projects. Held in Sissach, Switzerland, the conference was on the theme 'Towards a Sustainable Agriculture'. The programme had about 25 speakers from around the world, most prominently including Balfour. The proceedings volume (Besson and Vogtmann, 1978) covers a broad range of topics including soil fertility, livestock husbandry, pest management, biodynamic agriculture, plant breeding, economics, energy and nutrition. IFOAM's scientific conferences have been held every 2 or 3 years since then, the most recent (the 15th) having taken place in Adelaide, Australia, in 2005.

4.3.6 Ludwig Boltzmann Institute for Organic Agriculture and Applied Ecology, Vienna, Austria, 1980

The Ludwig Boltzmann Institute (available at: www.natur-wien.at/partner/boltzmann) was established in 1980 primarily to elaborate the scientific foundations of organic methods through interdisciplinary research. It is a private not-for-profit organization that is involved in educational activities at various universities, including advising on dissertations. Major areas of interest include plant production, composting, agroecology and food quality. The institute has an especially strong involvement in environmental protection in Vienna, particularly regarding rare and endangered species of plants and animals.

4.3.7 Elm Farm Research Centre, Hampstead Marshall, England, 1980

The Organic Research Centre (available at: www.efrc.com) was established in 1980 to address the major issues raised by a global economy based on intensive agriculture. It is a charitable trust based at Elm Farm, a 94 ha organic farm, and also works with a network of established organic farms. It is the UK's leading research, development and advisory institution for organic agriculture, and has been important in the development of organic research, policy and standards.

4.3.8 First chair of organic farming, Witzenhausen, Germany, 1981

In 1981, Hartmut Vogtmann, who at the time was director of FiBL, was invited to assume the world's first professorship in organic agriculture, established at Witzenhausen at the Gesamthochschule Kassel (Comprehensive University of Kassel, now called University of Kassel). This was a major milestone in the history of organic farming research. A 'conversion' of the whole faculty to a 'Division of Organic Agricultural Sciences' (available at: www.uni-kassel.de/fb11cms/) occurred in 1997. The professorship in Witzenhausen was followed by another chair for organic farming in 1987 at the Institute for Organic Farming of the University of Bonn. Throughout the 1990s, several professorships and coordination posts at institutes of higher education were established in Germany (Eberswalde, Giessen, Kiel, Munich, Nuertingen, Osnabrueck, Stuttgart-Hohenheim, Wiesbaden-Geisenheim) (Haccius and Lünzer, 2000).

4.3.9 DARCOF, Denmark, 1995

The Danish Research Centre for Organic Food and Farming (DARCOF) (available at: www.darcof.dk) was established in 1995 as a 'centre without walls', with the actual research performed in interdisciplinary collaborations among the participating research groups. The mission of DARCOF is to coordinate research for organic farming, with a view to achieving optimum benefit from the allocated resources. It seeks to elucidate the ideas and problems faced in organic farming by promoting high-quality research meeting international standards. During the first years of DARCOF's existence several unique research facilities were set up to provide an opportunity for conducting different projects simultaneously, using the same research fields, herds, etc. This allows close cooperation among different research environments, with a high degree of interdisciplinary collaboration, synergy and complementary research.

4.3.10 European Union funding

A wide range of research projects have been funded under the European Union's (EU's) Framework Programmes since the 1990s, as well as under various national programmes, some of which go back further. This research is carried out at a great many universities and other research institutions throughout Europe. More details are given in Chapter 6.

4.3.11 Organic research programmes at US Land Grant universities

The Sustainable Agriculture Farming Systems (SAFS) project was established in 1988 at the University of California, Davis, to study alternative agricultural systems using an interdisciplinary approach. The first phase of the SAFS project, completed in 2000, focused on agronomic differences among conventional, low-input and organic systems. SAFS established itself as a leader in agroecosystem research and education projects that quantify and analyse complex ecological and economic consequences of the transition from conventional to non-conventional farming systems.

Several additional programmes emerged in the late 1990s at various Land Grant universities (the universities that are the homes of a nationwide system of public agricultural colleges), including North Carolina State University's Center for Integrated Farming Systems, Ohio State University's Organic Food and Farming Education and Research

Program, West Virginia University's Organic Research Farm, and the University of Minnesota's Southwest Research and Outreach Center. Iowa State University was the first US Land Grant university to establish an organic extension position (1997).

4.4 The Modern Era

From its small beginnings, largely on the fringe of agricultural science, organic research has grown rapidly, especially since the mid-1990s. This section highlights a few topics that have been particularly prominent over the years, with a few examples of research in each. Most of these topics were already important in the thinking of the organic pioneers. The point of this summary is not to compile everything we have learned in these topics, but rather to give a sense of what areas organic researchers have concentrated on and the kinds of research they have done. The emphasis is on research that elucidates the principles and mechanisms of organic farming, i.e. that helps explain how and why organic farms perform the way they do, rather than simply measuring how well they perform.

4.4.1 Soil ecology

Studies emphasizing various aspects of soil ecology, including soil physical, chemical and biological properties and soil ecological processes, such as nutrient cycling, represent a large proportion of the scientific literature on organic farming. This is to be expected, given the considerable effort needed to maintain adequate levels of available nutrients for crop production without the addition of commercial fertilizers, and with reliance on organic sources of nutrients and associated cycling processes. As noted, a major assertion of organic farming proponents historically and today is that this system promotes soil health by maintaining high levels of fresh organic matter and biological activity.

The modern organic farming literature has many examples of studies that compare various soil ecological variables in organic and conventional farming systems and confirm the consensus that early proponents were correct – organic farming does enhance overall soil quality and health, by many measures. The various long-term farming systems experiments in Europe and the USA have been particularly important for validating the pioneers' early claims.

Some of the most significant findings come from the Swiss 'DOK' (biodynamic, organic and conventional) trial (see Chapter 14, this volume). Siegrist *et al.* (1998) found significantly greater aggregate stability, earthworm biomass, density and population diversity in the organic plots. In agreement with Howard's claim (1947), Mäder et al. (2000) found that mycorrhizal fungal colonization was 30–60% higher in plants grown in soils from the organic than in conventional farming systems, and concluded that organic farming systems had a greatly enhanced capacity to initiate fungal–plant symbiosis. Fließbach and Mäder (2000) showed that microbial biomass C and N, as well as their ratios to the total and light fraction C and N pools, were higher in soils in the organic systems than the conventional ones. They interpreted this as indicating enhanced decomposition of the easily available light fraction pool of SOM with increasing amounts of microbial biomass. In addition, Fließbach et al. (2000) found that the biodynamic system had a soil microbial community that was more efficient in using substrates for growth than the other systems in a ^{14}C-labelled decomposition study. Recently, Mäder et al. (2002), summarizing various characteristics of these systems after 21 years, reported that the organic systems were higher in soil aggregate stability, increased microbial biomass, dehydrogenase, protease, phosphatase, root length colonized by mycorrhizae, earthworm biomass and abundance, and density of carabids and staphylinid beetles (Coleoptera) and spiders (Arachnida). Furthermore, increased microbial diversity associated with decreased metabolic quotient in the organic systems was thought to indicate that the organic systems are more efficient at resource utilization, a characteristic of mature ecosystems (Mäder et al., 2002).

In the USA, studies conducted in the Rodale Institute's Farming Systems Trial (FST) have provided similar support regarding several soil ecological variables. Doran et al. (1987) showed that soil microbial biomass levels and reserves of potentially mineralizable N were greatest in the two alternative systems that had legumes. Significantly higher levels of microbial biomass, fungi, bacteria, dehydrogenase enzyme activity, soil bulk density and soil respiration were found in the surface soil layer in the manure-based organic system than in the conventional maize–soybean system (Doran et al., 1987). Werner and Dindal (1990) found that high levels of CO_2 evolution (a measure of potential microbial activity) in the organic plots correlated with high inputs of organic matter, and that soil nematodes were most abundant in organic plots. After 8 years, Wander et al. (1994) found that while changes in SOM content of the FST soils were still small, there were important changes in the biologically active and more stable, but still labile components of SOM, suggesting that particulate organic matter (measured as the light fraction) is functionally important in organic systems. About 15 years after conversion, Drinkwater et al. (1998) found that the legume-based organic system had greater net balances of both C and N than did the conventional maize–soybean system, and attributed this difference to the use of low C/N organic residues to maintain soil fertility, combined with greater temporal diversity of cropping sequences in the organic legume-based system.

By far, the majority of published studies on nutrient cycling in organic farming emphasize N. Although it has been assumed that N availability limits crop productivity under organic management, this pattern has not been generally supported by empirical studies, especially after a transition period of 5–7 years. To address both the short-term and long-term consequences of organic farming, several studies have developed farm and enterprise nutrient budgets. Although some authors had predicted that organic agriculture would deplete soil nutrients over time (e.g. Magid *et al.*, 1995), empirical studies support the opposite conclusion. For example, Watson *et al.* (2002) published a review of nutrient budgeting for organic farms representing temperate areas, and concluded that on an average, organic farms had positive balances for N, P and K, although there was considerable variation in nutrient use efficiency (outputs/inputs).

4.4.2 Nutrient losses

Uptake of nutrients by microbial biomass through immobilization is important in controlling leaching of nutrients from agroecosystems during parts of the year when there is no crop uptake. Numerous studies have shown lower leaching of nitrates from organic farms than conventional ones. For example, Eltun *et al.* (1995) compared N leaching in ecological and conventional cropping systems in an experiment in Norway and found that the nitrate runoff in the conventional cash crop system was more than twice as high as in the ecological cash crop system. For the forage crop systems, the nitrate loss in the ecological system was reduced by 36% compared with the conventional system. The most important factors influencing the N runoff in the different cropping systems seemed to be crop rotation, soil tillage, time of manure application and amount of fertilizer. However, Scheller and Vogtmann (1995) conducted on-farm case studies of N-mineralization on organic farms in Europe and found several causes of high accumulation of leachable nitrate in soils: fallow during the growing season, fallow during autumn and winter, manuring in autumn, leaching of chopped green manures by rain, low N uptake of crops because of diseases and transfer of mineralizable N from spring to autumn.

Kristensen *et al.* (1994) compared the leachable inorganic N content in soils from 26 organic and 550 conventional farms in Denmark during autumn 1990 and found the average nitrate-N content (0–75 cm) was similar in organic (31 kg/ha) and conventional farms using manure

(29 kg/ha), but lower in conventional farms not using manure (22 kg/ha). There was no significant increase in nitrate leaching risk in organic farms compared with conventional farms applying manures; the differences in nitrate levels appeared to be related to the use of manures as opposed to inorganic fertilizers. Following up on this earlier study, using representative data, Knudsen et al. (2006) found a lower N-leaching loss from organic farms than from conventional mixed dairy farms, primarily due to lower N inputs. The N-leaching loss depended on soil type, the use of catch crops and the level of SOM, and was highest on sandy soils with high SOM and no catch crops. The authors stressed the importance of using representative data from organic and conventional farming practices in comparative studies of N-leaching loss, and stated that lack of representative data has been a major weakness of previous comparisons.

4.4.3 Natural controls of insect pests and diseases

Since organic farmers do not use most pesticides, an obvious question is: are pests kept to acceptable levels, and if so, how? Research since the early 1990s has shown that some insect pests can cause problems in some organic crops. However, even more studies report lower populations of insect pests or no difference between organic and conventional farms.

Predation and parasitization are two mechanisms for natural control of insect pests. Several researchers have looked at the abundance and diversity of naturally occurring beneficial insects and other arthropods. Beginning around 1980 and continuing to the present, numerous researchers working in diverse crops around the world have found that organic farms host a wider range of beneficial species of arthropods than do conventional farms. Categories of natural enemies and other beneficial species found to be higher on organic farms include: carabid beetles, staphylinid beetles, spiders, parasitoids, non-parasitic nematodes, dung beetles and non-pest butterflies. Later papers have shown the importance of semi-natural habitat in combination with organic farming for supporting healthy populations and a high diversity of carabids and epigeal spiders (Pfiffner and Luka, 2000, 2003).

An important group of predators that has received considerable attention in organic research is ground beetles (Coleoptera: Carabidae). Several studies from 1980 to 2005 report greater abundance of many species of ground beetles in organic than in conventional fields. In the earliest reported study (Dritschilo and Wanner, 1980), populations were from 20% to almost 700% higher in organic farms in the midwestern USA. The organic farms also had about twice the number of species found on conventional farms, but had approximately the same level of diversity as measured by the Shannon–Wiener index. However, based on his studies of carabids in seed potato fields in Scotland, Armstrong (1995) concluded that organic management does not necessarily bring greater diversity and abundance.

A landmark study published in 1990 of earthworms and soil microarthropods (nematodes, fungivorous Prostigmata mites, oribatid mites, predatory Mesostigmata mites and Collembola) conducted in the Rodale Farming Systems Trial points out both the benefits and negative consequences of organic farming for these organisms (Werner and Dindal, 1990). The authors concluded that while organic amendments tend to enhance soil biological activity at times during the yearly cycle, tillage tends to disrupt the biotic community.

A comprehensive analysis that significantly contributed to our understanding of how organic farms function with respect to insect pests and natural enemies was that of Letourneau and Goldstein (2001), who investigated pest damage and the arthropod community structure on organic and conventional tomato farms in California. They found no difference in herbivore abundance but higher natural enemy abundance and greater species richness of all functional groups of arthropods (herbivores, predators, parasitoids, etc.) on the organic farms. The authors concluded that on the organic farms, any particular pest species would have been diluted by a greater variety of herbivore species and would be subject, on average, to a wider variety and greater abundance of potential parasitoids and predators.

4.4.4 Crop resistance to pests and diseases

Early proponents of organic farming such as Howard and Balfour argued that their methods produce 'healthy' crops that are less susceptible to insects and diseases. Several researchers have since found that organic fertilization has a positive effect on the resistance of plants to insects and disease.

Two papers of particular importance from the mid-1990s lend credence to the pioneers' view. Phelan et al. (1995) and Phelan (1997) investigated mechanisms that could explain why insect pests can be lower on organic farms. Soils from organic and conventional farms were brought into the laboratory and treated with various conventional and organic amendments. Maize plants were grown in pots of these soils in the greenhouse and exposed to female European corn borers (ECB), Ostrinia nubilalis. The insects consistently preferred to lay eggs on plants grown in soils with a history of conventional management no matter what soil amendment was applied. Subsequent studies suggested that differences in ECB ovipositional preference were related to the plant–mineral balance. The ability of organically managed soils to buffer nutrient uptake was also demonstrated by an analysis of profiles of eight minerals in maize plants grown in soils from organic and conventional farms (Phelan, 1997). When compost or NH_4NO_3 was added to the organic soil, the plant mineral profile showed little change compared with unfertilized plants, but the plants in the conventionally managed soil showed dramatic shifts, particularly when amended with NH_4NO_3. Furthermore, variability in the data was greater in conventional than in organic soils, suggesting that a more resistant physiological state is more likely in organically managed soils because of the inherently greater capacity of these soils to buffer the availability of minerals to plants.

With respect to plant diseases, many researchers have found lower incidences of some diseases on organic farms, such as Fusarium head blight (Oerke et al., 2001), owing to long rotations, compost additions and other management factors. However, others have found variable or greater disease incidence in various crops under organic management than under conventional management. For example, Baturo (2002) found the worst plant health status of spring barley in organic farming systems, especially in samples of the stem base. In that study, some *Fusarium* spp. were more abundant in the organic system, while others were more abundant in the conventional system.

4.4.5 Crop and food quality

The pioneers' belief that organic farming would produce healthier foods than conventional farming because they are grown in healthier soils continues to be a debated issue. In 1983, Rasmussen (1983) found that crops on a long-time biodynamic farm in Denmark had lower N and Ca and higher K uptakes than those on a conventional farm. Furthermore, the nutritive value for rats was higher in barley from the biodynamic farm, but for wheat it did not differ between farms (Rasmussen, 1983). In a review of the European literature, Vogtmann (1984) reported that many researchers had found that nitrate levels were lower in vegetables grown with composted farm manure compared with mineral fertilizers, that organically fertilized vegetables often had higher levels of desirable components such as vitamin C, various trace elements, iron and β-carotene, and that potatoes and vegetables from biological systems kept better than those receiving mineral fertilizers. However, Vogtmann (1984) stated that the influence of agricultural practices on the composition of plant foods and its consequences for human health could not be demonstrated conclusively with existing data at that time.

In one of the few studies that directly links soil fertility to mineral status of a crop, Garcia *et al.* (1989) got variable results with avocados from biodynamic and conventional systems in Tenerife. SOM, pH and available P, Ca, Mg and K were all greater in the biodynamic soils, while N, P, K, Mg and Cu plant tissue levels were not different; Ca and Mn levels were lower, and only Zn was higher in biodynamic plants. Recently, Reeve *et al.* (2005) in a 6-year study in the USA found that biodynamic management enhanced wine grape quality, but found no significant differences in soil quality compared with conventional management.

Velimirov *et al.* (1992) conducted one of the few studies on the influence of organically and conventionally grown food on the fertility of rats; their work echoes some of the earliest studies conducted by McCarrison in the early 20th century (see Chapter 2, this volume). Biologically and conventionally grown products of the same variety, obtained from neighbouring farms in Austria, were compared for their influence on the fertility of two groups of laboratory rats up to the third generation. There was no significant difference in the pregnancy rate between the two groups. The average litter weight was mostly higher in the biological than conventional group, but not significantly so. There were significantly fewer perinatally dead offspring in the biologically fed group. The biologically fed females had a much greater ability to compensate for weight loss during and after lactation, and their weight gain was significantly higher than in the conventionally fed group.

Woese *et al.* (1997) conducted another review of the food quality literature and had difficulty drawing clear conclusions because different methods of sampling were used in the investigations. However, they were able to make the following few generalizations. Organic foods had lower pesticide and nitrate levels than conventional foods. In the case of leafy vegetables, a higher dry matter concentration was observed in organically grown or fertilized products than in conventionally or minerally fertilized products. In feed selection experiments, the animals (rabbits, mice, rats, hens and pigeons) preferred organic produce. However, conventional wheat, with its higher protein and superior protein quality, did better in meeting common baking requirements.

Finally, in an excellent example of a recent highly controlled study (Kihlberg *et al.*, 2006), wheat grown in organic and conventional experimental fields was milled and baked for sensory evaluation for over 2 years. The differences between years dominated differences between farming systems, except that bread made from conventional flour had significantly higher elasticity and moistness.

4.4.6 Weed ecology

Weeds are a major component of most organic farming systems. The generally lower yields reported in organic farming compared with conventional farming are often attributed to crop–weed competition. Furthermore, organic weed management strategies, particularly with intensive mechanical cultivation, are major forces determining the structure and function of organic farming systems and have important effects – often undesirable – on soil ecology and quality (Chapter 5, this volume, describes the considerable progress in organic weed control techniques over the past several decades).

Not surprisingly, several researchers have found higher weed densities and greater weed biomass on organic farms compared with conventional farms. However, Ngouajio and McGiffen (2002) reported that weed density and biomass were often lower in organic systems, and cited weed seed and seedling predation and physical and allelopathic effects of cover crops as factors that may suppress weeds in organic agriculture.

Numerous researchers have found organic cropping practices to be beneficial for species diversity of weed communities compared with conventionally cropped fields. In addition to greater overall species diversity of weeds, considerably higher numbers of rare and endangered weed taxa have been reported on decades-long organic farms compared with conventional farms, leading to suggestions that organic farming may contribute to maintaining plant biodiversity in agricultural landscapes. However, although some of the early studies found at least twice as many weed species on organic compared with conventional farms, Hyvonen *et al.* (2003) suggested that these studies may overestimate the benefits of organic cropping for a number of weed species because of the higher number of individuals found in organic than conventional fields. In their study, which compared newly converted organic farms with low-intensity conventional farms in Finland, when mean weed species number was adjusted for number of individuals they found much smaller differences.

In contrast to on-farm studies showing increased numbers of rare and endangered plant species on decades-old organic farms, Albrecht and Matteis (1998) did not find that converting to organic management significantly increased these species in the first 4 years of their study in southern Bavaria. However, rare species showed roughly constant frequencies and densities in the organic system, and there was a positive correlation between the number of rare species and the total weed cover, suggesting that reducing management intensity might have led to an increase (Albrecht and Mattheis, 1998).

There is evidence that greater weed abundance in organic farming provides favourable habitat for more abundant and diverse populations of carabid beetles (Coleoptera: Carabidae), including endangered stenoceous carabids (Pfiffner and Luka, 2003). In addition, greater weed diversity on organic farms was found to influence epigeal spider fauna (Pfiffner and Luka, 2003). These studies have contributed to the claim that organic farming in combination with semi-natural habitats may be an important factor for the conservation and enhancement of general biodiversity on agricultural landscapes (Rydberg and Milberg, 2000; Pfiffner and Luka, 2003).

4.4.7 Livestock[1]

Research to improve and develop organic animal husbandry started later and has been less extensive than research dealing with crop production. There are several explanations, for example, that animal research typically requires more resources than crop research. Much of the early organic research was done on farms and by farmers, and it was cheaper and easier to experiment with crops than to design livestock trials. The universities, which had the necessary research facilities, were slow to follow the organic trend in animal research. At the first IFOAM Scientific Conference in 1977, described earlier, only one of the 25 contributions in the proceedings volume (Besson and Vogtmann, 1978) was about animal husbandry. The next time the IFOAM conference was held in Switzerland (in Basel), 23 years later, the number of contributions was up more than 20-fold, to well over 500. However, the proportion dealing with animal production was only about 8% (Alföldi et al., 2000).

This lack of emphasis on animal husbandry seems to be true not only for research, but also for production. Organic animal husbandry in general has developed more slowly than organic plant production, including the development of standards. For example, the EU regulations on organic livestock production came 8 years after the regulations for plant production (see Chapter 8, this volume). In part, this may have had to do with the underlying philosophy of organic farming. The organic movement has generally developed primarily from environmental concerns rather than animal welfare concerns, e.g. with an early focus on soil fertility and human health. As a result, it has been more difficult to define the essence (principles) of organic livestock production (Lund, 2002), and without these principles it has been difficult to agree on guidelines and standards.

---
[1]This section was written by Vonne Lund, National Veterinary Institute, Oslo.

Even so, there has been some influential organic research in the field of animal husbandry.

Initially, most of the animal research was performed in northern Europe, particularly in the Scandinavian countries and Switzerland, with overseas research almost non-existent. An important European initiative was the establishment of an EU-network, Network for Animal Health and Welfare in Organic Agriculture (NAHWOA, 1999–2001; available at: www.veeru.reading.ac.uk/organic). This brought researchers together from all over Europe, establishing a broad arena for discussion and inspiration that perhaps helped participants less familiar with organic thinking to 'go organic'. Also, for several years, the proceedings from the network (published online) were the best available source of research reports. The NAHWOA was followed by another EU-financed network, 'Sustaining Animal Health and Food Safety in Organic Farming' (SAFO, 2003–2006; available at: www.safonetwork.org), again an important meeting arena for animal scientists interested in organic farming.

Organic production is supposed to be based on local resources (including local and indigenous knowledge). Thus, 'best production practice' in organic farming depends on local conditions and varies among countries and perhaps regions. The development of organic farming is not only a result of different climatic and geographical conditions, but is also highly dependent on the prevailing institutional and political framework. Across Europe, this has resulted in a diversity of livestock systems (Roderick et al., 2004). Research needs have varied accordingly. The research that has been performed has to a large extent been very applied, since it has grown out of the practical needs of the farmers, which also means that its focus has differed from country to country. An example from two neighbouring countries, Denmark and Sweden, will illustrate this. In Denmark, research to develop farming systems based on clover–grass production, including the use of large proportions of roughage (hay and silage) in cattle rations, is pointed out as one of the most significant contributions of organic farming research. It is seen as having had an important impact on conventional farming systems, as well as being fundamental for the development of organic farming systems. In contrast, in Sweden, this type of research is rarely mentioned as an achievement of organic farming, since mixed farming systems with a significant proportion of clover–grass leys in the rotation never went out of fashion in conventional farming, at least not in the less fertile agricultural areas of Sweden.

As for breaking new ground and changing thought patterns, an important contribution of organic livestock research emphasizes preventative health (in a wide sense) rather than curing diseases. Using a systems approach, that is, seeing the animal and its health in its total environment and as part of the farming system, the aim has been to minimize stress and other negative effects on the animal, thereby enhancing its immune system and optimizing its disease resistance (e.g. Boehncke, 1986). An important part of this research has focused on minimizing the use of antibiotics. The need for this kind of research has been further underlined by the fact that antibiotics may have a harmful impact on the ecosystem (e.g. Waller, 1997; Sangster, 1999). Several organic research projects have therefore focused on disease prevention. Research has also examined alternative treatments such as homoeopathy. Although the results with homoeopathic treatments so far are mixed, several projects show that mastitis may be treated successfully without antibiotics (e.g. Hektoen, 2004; Notz et al., 2005; Klocke et al., 2006). The EU regulation on organic animal production recommends homoeopathy and other non-allopathic treatment methods to treat sick animals, rather than conventional veterinary medicine. In Switzerland, the pro-Q project provides an example of how to change the focus from treating diseases to keeping animals healthy on organic farms. The aims of the project are reduction of the use of antibiotics in udder treatment, improvement in the udder health status of the herds and, as a consequence, improvement in milk quality.

Another central theme that has influenced conventional animal production is the emphasis on naturalness and natural behaviour (Vaarst et al., 2001; Lund, 2002). In organic farming the animal welfare concept is interpreted in terms of natural living, which includes providing the animal with feeds adapted to its physiology, the possibility of performing its natural behaviour and a natural environment. This has led to the development of alternative rearing systems aimed at allowing animals to have a more natural life (Roderick et al., 2004). Many of the issues that from the start have been close at heart to the organic movement are now considered as goals or are even being implemented in conventional animal production, particularly in the EU. The EU in 2006 placed a ban on tethering of sows and required that they have rooting materials and increased living space. Furthermore, veal calves may not be crated after 8 weeks of age, and conventional cages for poultry will be banned from 2012. The organic approach also puts the focus on breeding and breeding goals as an important means to achieve optimal yield, increased disease resistance and better mothering abilities that make the animals better adapted to free range and group housing conditions. (For traits with low heritability, however, there is an inherent conflict between the aim of natural living and the aim of efficient breeding, such as for disease resistance, since the latter requires big groups of offspring and thereby the use of artificial insemination.)

Based on the idea of natural living, organic farmers have persisted in feeding their cows a 'natural' diet based on roughage, in spite of conventional advisers warning against this practice, arguing that it would cause severe malnutrition in today's high-yielding dairy breeds. However, research projects in Norway, for example, showed that organic dairy cows did not display increased frequency of ketosis; rather the opposite was true (e.g. Hardeng and Edge, 2001; Hamilton *et al.*, 2002; Bennedsgaard *et al.*, 2003). The 'secret' behind the organic success was that the farmers were consistent in their feeding strategy and started feeding roughage to calves and heifers at an early age, which allowed the rumen to develop properly. Eventually the findings led to feeding recommendations being modified in conventional farming also. It has also been shown that the frequency of mastitis is lower in Swedish organic herds, but the mechanisms behind this have not yet been clarified (Hamilton *et al.*, 2006).

In summary, although research on organic animal production got off to a late start, the situation has changed markedly in the last decade. Animal research now is a significant part of the overall picture, and has significantly contributed not only to organic production but also to recent thinking about conventional systems.

4.5 Conclusion

Organic farming research began as a strong reaction against the mechanistic and reductionistic impulse that emerged in agricultural science in the early and mid-20th century and that still dominates. The roots of the continuing debate between conventional agriculture on one hand and sustainable, alternative or organic agriculture on the other lie in deeply imbedded opposing world views. There are many who say that contemporary organic research has already lost its holistic roots and is going the way of reductionistic science as organic foods become part of the mainstream market. However, there is now a creative potential between the two camps that could, and indeed in some cases is, leading to cracks in the disciplinary armour in several scientific agricultural institutions. For example, many US Land Grant universities have developed some type of interdisciplinary 'systems' programmes in agriculture, whether under the banner of 'sustainable' or organic agriculture or 'agroecosystems management'. Many of these programmes show the influence of the ideas of the early organic pioneers such as Howard and Balfour, although not necessarily explicitly. As these programmes grow and expand in influence within their institutions, the institutions themselves undergo change.

Will the mainstream institutions ever fully embrace the holistic/systems ideas of the organic pioneers? Would this then bring utopia to agriculture, the oldest way that humans mould their environment to suit their needs? Both holistic and reductionistic approaches are needed to advance the efficiency of organic farming. Ideally, mechanistic or reductionistic studies will be conducted in a larger systems context. In that way, the best of both approaches can be combined. However, organic farming, in particular – as should be the case for all farming – involves both science and art. The art comes from the farmers who live with the land and practice organic farming in its full systems context. It is based on their experiential knowledge and insights on what works or might work on their farms and in the larger community. The tools of science can be brought to bear to refine and make ideas inspired by the art of farming useful to a wider audience. In this way, a spiral of co-learning between scientists and farmers is created that will synergistically advance the science and praxis of organic farming far into the future.

Acknowledgement

A tragic car accident prevented the late Benjamin R. Stinner, one of the earliest US scientific leaders to study organic farming, from co-authoring this chapter as planned. Nevertheless, he had considerable direct input in its ideas and outline. He has also had significant indirect input through the many years of lively discussions on the science of organic farming that the two of us shared.

References

Albrecht, H. and Mattheis, A. (1998) The effects of organic and integrated farming on rare arable weeds on the Forschungsverbund Agrarökosysteme München (FAM) research station in southern Bavaria. *Biological Conservation* 86, 347–356.

Alföldi, T., Lockeretz, W. and Niggli, U. (eds) (2000) *IFOAM 2000 – The World Grows Organic*. Proceedings of the 13th International IFOAM Scientific Conference, Basel, 28–31 August 2000. vdf Hochschulverlag AG an der ETH Zürich, Switzerland.

Armstrong, G. (1995) Carabid beetle (Coleoptera, Carabidae) diversity and abundance in organic potatoes and conventionally grown seed potatoes in the north of Scotland. *Pedobiologia* 39, 231–237.

Baars, T. (2002) Reconciling scientific approaches for organic farming research: Part I: Reflection on research methods in organic grassland and animal production at the Louis Bolk Institute, The Netherlands. PhD thesis, Department of Livestock Production, Louis Bolk Instituut. Published in Louis Bolk Instituut Publications no. G38. Louis Bolk Instituut, Driebergen, Utrecht.

Balfour, E.B. (1948) *The Living Soil*. Faber & Faber, London.

Balfour, E.B. (1978) The living soil. In: Besson, J.-M. and Vogtmann, H. (eds) *Towards a Sustainable Agriculture*. Verlag Wirz AG, Aarau, Switzerland, pp.18–27.

Barton, G. (2001) Sir Albert Howard and the forestry roots of the organic farming movement. *Agricultural History* 75, 168–187.

Baturo, A. (2002) Health status of spring barley cultivated under organic, integrated and conventional farming conditions. In: *The BCPC Conference: Pests and Diseases.* Proceedings of an International Conference. British Crop Protection Council, Farnham, UK, pp. 699–704.

Bennedsgaard, T.W., Thamsborg, S.M., Vaarst, M. and Enevoldsen, C. (2003) Eleven years with organic dairy production in Denmark – herd health and production related to time of conversion and compared to conventional production. *Livestock Production Science* 80, 121–131.

Besson, J.-M. and Vogtmann, H. (eds) (1978) *Towards a Sustainable Agriculture.* Verlag Wirz AG, Aarau, Switzerland.

Boehncke, E. (1986) Die Auswirkungen intensiver Tierproduktion auf das Tier, den Menschen und die Umwelt. In: Sambraus, H.H. and Boehncke, E. (eds) (1986) *Ökologische Tierhaltung.* Alternative Konzepte 53. Verlag C.F. Müller, Karlsruhe, Germany, pp. 9–26.

Conford, P. (1995) The alchemy of waste: the impact of Asian farming on the British organic movement. *Rural History* 6, 103–114.

DeGregori, T.R. (2004) *Origins of the Organic Agriculture Debate.* Iowa State Press, Ames, Iowa.

Delate, K. (2002) Using an agroecological approach to farming systems research. *HortTechnology* 12, 345–354.

Doran, J.W., Fraser, D.G., Culik, M.N. and Liebhardt, W.C. (1987) Influence of alternative and conventional agricultural management on soil microbial processes and nitrogen availability. *American Journal of Alternative Agriculture* 2, 99–106.

Drinkwater, L.E., Wagoner, P. and Sarrantonio, M. (1998) Legume-based cropping systems have reduced carbon and nitrogen losses. *Nature* 396, 262–265.

Dritschilo, W. and Wanner, D. (1980) Ground beetle abundance in organic and conventional corn fields. *Environmental Entomology* 9, 629–631.

Eltun, R., Kristensen, L., Stopes, C., Kølster, P., Granstedt, A. and Hodges, D. (1995) Comparisons of nitrogen leaching in ecological and conventional cropping systems. In: Kristensen, L. (ed.) *Nitrogen Leaching in Ecological Agriculture.* Proceedings of an International Workshop, Royal Veterinary and Agricultural University, Copenhagen, Denmark. AB Academic Publishers, Bicester, UK, pp. 103–114.

Fließbach, A. and Mäder, P. (2000) Microbial biomass and size-density fractions differ between soils of organic and conventional agricultural systems. *Soil Biology and Biochemistry* 32, 757–768.

Fließbach, A., Mäder, P. and Niggli, U. (2000) Mineralization and microbial assimilation of ^{14}C-labeled straw in soils of organic and conventional agricultural systems. *Soil Biology and Biochemistry* 32, 1131–1139.

Garcia, C., Alvarez, C.E., Carracedo, A. and Iglesias, E. (1989) Soil fertility and mineral nutrition of a biodynamic avocado plantation in Tenerife. *Biological Agriculture and Horticulture* 6, 1–10.

Haccius, M. and Lünzer, I. (2000) Organic Agriculture in Germany. Stifung Ökologie & Landbau (SÖL), Bad Dükheim, Germany. Available at: http://www.organic-europe.net/country_reports/pdf/2000/germany.pdf

Hamilton, C., Hansson, I., Ekman, T., Emanuelson, U. and Forslund, K. (2002) Health of cows, calves and young stock on 26 organic dairy herds in Sweden. *Veterinary Record* 150, 503–508.

Hamilton, C., Emanuelson, U., Forslund, K., Hansson, I. and Ekman, T. (2006) Mastitis and related management factors in certified organic dairy herds in Sweden. *Acta Veterinaria Scandinavica* 48, 11.

Hardeng, F. and Edge, V.L. (2001) Mastitis, ketosis, and milk fever in 31 organic and 93 conventional Norwegian dairy herds. *Journal of Dairy Science* 84, 2673–2679.

Hektoen, L. (2004) Homeopathic treatment of farm animals: studies of utilisation, effects and implications for animal health and welfare. PhD thesis. Department of Production Animal Clinical Sciences, Norwegian School of Veterinary Science, Oslo, Norway.

Howard, A. (1943) *An Agricultural Testament*. Oxford University Press, Oxford. Originally published in 1940.

Howard, A. (1945) *Farming and Gardening for Health or Disease*. Faber & Faber, London.

Howard, A. (1947) *The Soil and Health: A Study of Organic Agriculture*. Devin-Adair, New York.

Howard, A. and Wad, Y.D. (1931) *The Waste Products of Agriculture: Their Utilization as Humus*. Oxford University Press, Oxford.

Hyvonen, T., Ketoja, E., Salonen, J., Jali, H. and Tianinen, J. (2003) Weed species diversity and community composition in organic and conventional cropping of spring cereals. *Agriculture, Ecosystems and Environment* 97, 131–149.

Jukes, T.H. (1981a) Organic farming. *Science* 213, 708.

Jukes, T.H. (1981b) Organic farming and the organic food concept. *Feedstuffs* 53, 26–29.

Kihlberg, I., Ostrom, A., Johansson, L. and Risvik, E. (2006) Sensory qualities of plain white pan bread: influence of farming system, year of harvest and baking technique. *Journal of Cereal Science* 43, 15–30.

Klocke, P., Ivemeyer, S., Walkenhorst, M., Maeschli, A. and Heil, F. (2006) Handling the dry-off problem in organic dairy herds by teat sealing or homeopathy compared to therapy omission. Paper presented at Joint Organic Congress, Odense, Denmark, 30–31 May 2006.

Knudsen, M.T., Kristensen, I.B.S, Berntsen, J., Petersen, B.M. and Kristensen, E.S. (2006) Estimated N leaching losses for organic and conventional farming in Denmark. *Journal of Agricultural Science* 144, 135–149.

Kolisko, E. and Kolisko, L. (1939) *Die Landwirtschaft der Zukunft*. Published in English as *Agriculture of Tomorrow* (1978). Kolisko Archive Publications, Bournemouth, UK.

Kristensen, S.P., Mathiasen, J., Lassen, J., Madsen, H.B. and Reenberg, A. (1994) A comparison of the leachable inorganic nitrogen content in organic and conventional farming systems. *Acta Agriculturae Scandinavica. Section B, Soil and Plant Science* 44, 19–27.

Letourneau, D.K. and Goldstein, B. (2001) Pest damage and arthropod community structure in organic vs. conventional tomato production in California. *Journal of Applied Ecology* 38, 557–570.

Liebig, J. von (1842) *Chemistry and Its Application to Agriculture and Physiology*. J. Monroe, Cambridge.

Lockeretz, W. (2000) Organic farming research, today and tomorrow. In: Alföldi, T., Lockeretz, W. and Niggli,

U. (eds) *The World Grows Organic*. Proceedings of the 13th IFOAM Scientific Conference, Basel 28–31 August 2000. vdf Hochschulverlag AG an der ETH Zürich, Switzerland, pp. 718–720.

Lund, V. (2002) Ethics and animal welfare in organic animal husbandry – an interdisciplinary approach. PhD thesis. Acta Universitatis Agriculturae Sueciae, Veterinaria 137. Department of Animal Environment and Health, Swedish University of Agricultural Sciences, Skara, Sweden.

Mäder, P., Edenhofer, S., Boller, T. Wiemken, A. and Niggli, U. (2000) Arbuscular mycorrhizae in a long-term field trial comparing low-input (organic, biological) and high-input (conventional) farming systems in a crop rotation. *Biology and Fertility of Soils* 31, 150–156.

Mäder, P., Fließbach, A., Dubois, D., Gunst, L., Padruot, F. and Niggli, U. (2002) Soil fertility and biodiversity in organic farming. *Science* 296, 1694–1697.

Magid, J. and Kølster, P. (1995) Modelling nitrogen cycling in an ecological crop rotation – an exploratory trial. In: Kristensen, L., Stopes, C., Kølster, P., Granstedt, A. and Hodges, D. (eds) *Nitrogen Leaching in Ecological Agriculture*. Proceedings of an International Workshop, Royal Veterinary and Agricultural University, Copenhagen, Denmark. AB Academic Publishers, Bicester, UK, pp. 77–87.

Ngouajio, M. and McGiffen, M.E. (2002) Going organic changes weed population dynamics. *HortTechnology* 12, 590–596.

Niggli, U. (1999) Holistic approaches in organic farming research and development: a general overview. In: Zanoli, R. and Krell, R. (eds) *Research Methodologies in Organic Farming*. Proceedings, Frick, Switzerland, 30 September–3 October 1998. Food and Agriculture Organization of the United Nations, Rome.

Nitrogen Leaching in Ecological Agriculture. Proceedings of an International Workshop, Royal Veterinary and Agricultural University, Copenhagen, Denmark. AB Academic Publishers, Bicester, UK, pp. 77–87.

Notz, C., Klocke, P. and Spranger, J. (2005) Bestandesbetreuung und antibiotikaminimiertes Tiergesundheitsmanagement (BAT). Paper presented at 1st Swiss Buiatrics Congress, 19–21 October 2005, Berne, Switzerland, pp. 26.

Oerke, E.-C., Meier, A., Steiner, U. and Dehne, H.W. (2001) Incidence and control of Fusarium head blight in intensive wheat production in western Germany. *Phytopathology* 91(Suppl. 6), S67.

Pfeiffer, E. (1936) *Sensitive Crystallization Processes*. Anthroposophic Press, New York.

Pfeiffer, E. (1938) *Bio-dynamic Farming and Gardening; Soil Fertility Renewal and Preservation*. Translated by F. Heckel. Anthroposophic Press/Rudolf Steiner Publishing, New York/London.

Pfeiffer, E. (1940) *Bio-dynamic Farming and Gardening; Soil Fertility, Renewal and Preservation*, 2nd edn. Translated by F. Heckel. Anthroposophic Press, New York.

Pfeiffer, E. (1943) *Bio-dynamic Farming and Gardening; Soil Fertility, Renewal and Preservation*, 3rd edn. Translated by F. Heckel. Anthroposophic Press, New York.

Pfeiffer, E. (1984) *Chromatography Applied to Quality Testing*. Anthroposophic Press, New York.

Pfiffner, L. and Luka, H. (2000) Overwintering of arthropods in soils of arable fields and adjacent semi-natural habits. *Agriculture, Ecosystems and Environment* 78, 215–222.

Pfiffner, L. and Luka, H. (2003) Effects of low-input farming systems on carabids and epigeal spiders – a paired farm approach. *Basic and Applied Ecology* 4, 117–127.

Phelan, P.L. (1997) Soil-management history and the role of plant mineral balance as a determinant of maize susceptibility to the European corn borer. *Biological Agriculture and Horticulture* 15, 25–34.

Phelan, P.L., Mason, J.F. and Stinner, B.R. (1995) Soil-fertility management and host preference by European corn borer, *Ostrinia nubilalis* (Hubner), on *Zea mays* L: a comparison of organic and conventional chemical farming. *Agriculture, Ecosystems and Environment* 56, 1–8.

Rasmussen, J. (1983) Comparisons between farming systems in Denmark 1982. *Rapport, Institutionen for Vaxtodling, Sveriges Lantbruksuniversitet* 124: 88.

Reeve, J.R., Carpenter-Boggs, L., Reganold, J.P., York, A.L., McGourty, G. and McCloskey, L.P. (2005) Soil and winegrape quality in biodynamically and organically managed vineyards. *American Journal of Enology and Viticulture* 56, 367–376.

Rodale, J.I. (1948) *The Organic Front*. Rodale Press, Emmaus, Pennsylvania.

Roderick, S., Henriksen, B., Trujillo, R.G., Bestman, M. and Walkenhorst, M. (2004) In: Vaarst, M., Roderick, S., Lund, V. and Lockeretz, W. (eds). *Animal Health and Welfare in Organic Agriculture*. CAB International, Wallingford, UK, pp. 29–56.

Rydberg, N. and Milberg, P. (2000) A survey of weeds in organic farming in Sweden. *Biological Agriculture and Horticulture* 18, 175–185.

Sangster, N.C. (1999) Anthelmintic resistance: past, present and future. *International Journal for Parasitology* 29, 115–124.

Scheller, E. and Vogtmann, H. (1995) Case studies on nitrate leaching in arable fields of organic farms. *Biological Agriculture and Horticulture* 11, 91–102.

Siegrist, S., Schaub, D., Pfiffner, L. and Mäder, P. (1998) Does organic agriculture reduce soil erodibility? The results of a long-term field study on loess in Switzerland. *Agriculture, Ecosystems and Environment* 69, 253–264.

Steiner, R. (1971) *Theosophy. An Introduction to the Supersensible Knowledge of the World and the Destination of Man*. Anthroposophic Press, Hudson, New York.

Steiner, R. (1974) *Agriculture*. Biodynamic Agriculture Association, London.

Vaarst, M., Alban, L., Mogensen, L., Thamsborg, S.M. and Kristensen, E.S. (2001) Health and welfare in Danish dairy cattle in the transition to organic production: problems, priorities and perspectives. *Journal of Agricultural and Environmental Ethics* 14, 367–390.

van Steensel, F., Phillipa, N., Bauer-Eden, H., Kenny, G., Campell, H., Ritchie, M., Macgregor, A.N., Koppenol, M., Blake, G. and Bacchus, P. (2002) *A Review of New Zealand and International Organic Land Management Research Relevant to Soil, Dairy Pasture and Orchard Management in New Zealand*. The Research and Development Group of the Bio Dynamic Farming and Gardening Association, New Zealand.

Velimirov, A., Plochberger, K., Huspeka, U. and Schott, W. (1992) The influence of biologically and conventionally cultivated food on the fertility of rats. *Biological Agriculture and Horticulture* 8, 325–337.

Vogtmann, H. (1984) Organic farming practices and research in Europe. In: Bezdicek. D.F., Power, J.F., Keeney, D.R. and Wright, M.J. (eds) *Organic Farming: Current Technology and Its*

Role in a Sustainable Agriculture. ASA Special Publication No. 46. American Society of Agronomy, Madison, Wisconsin, pp. 19–36.

Voisin, A. (1999) *Soil, Grass and Cancer*. Acres USA Publishers, Austin, Texas.

Waller, P.J. (1997) Anthelmintic resistance. *Veterinary Parasitology* 72, 391–412.

Wander, M.M., Traina, S.J., Stinner, B.R. and Peters, S.E. (1994) Organic and conventional management effects on biologically active soil organic matter pools. *Soil Science Society of America Journal* 58, 1130–1139.

Watson, C.A., Bengtsson, H. Ebbesvik, M., Løes, A.-K., Myrbeck, A., Salomon, E., Schroder, J. and Stockdale, E.A. (2002) A review of farm-scale nutrient budgets for organic farms as a tool for management of soil fertility. *Soil Use and Management* 18(Suppl.), 264–273.

Werner, M.R. and Dindal, D.L. (1990) Effects of conversion to organic agricultural practices on soil biota. *American Journal of Alternative Agriculture* 5, 24–32.

Woese, K., Lange, D., Boess, C. and Boegl, K.W. (1997) A comparison of organically and conventionally grown foods – results of a review of the relevant literature. *Journal of the Science of Food and Agriculture* 74, 281–293.

Woodward, L. (2002) Science and research in organic farming. EFRC pamphlet series. Policy and Research Department, Elm Farm Research Centre, Hampstead Marshall, UK.

5 The Evolution of Organic Practice

U. Niggli

Director, Research Institute of Organic Agriculture (FiBL), Ackerstrasse, 5070 Frick, Switzerland

5.1 The Rejection of Conventional Agricultural Techniques

Organic farming was initially characterized, in part, by a sometimes fierce rejection of the so-called conventional farming techniques characteristic of mainstream agriculture. In mainstream agriculture, crop production was first intensified by the use of commercial fertilizers. Surprisingly, both conventional scientists and organic pioneers invoked the same scenario of imminent soil depletion. The former were concerned about an insufficient supply of macro- and micronutrients, while the latter warned about a complete loss of biological soil fertility (see Chapter 2, this volume). Consequently, both groups saw the productivity of agriculture as endangered. The next step in the intensification of agriculture was the widespread use of insecticides, fungicides and herbicides, a practice that also made many conventional farmers feel uncomfortable.

The pursuit of yield increases also took hold in livestock husbandry, leading to changes in feeding regimes, industrialized methods for keeping animals and increasing use (and misuse) of allopathic medicine. The latest and most significant step in that direction has been the genetic engineering of crops, livestock and microorganisms.

On the basis of a general concept of 'naturalness', organic pioneers defined themselves largely by not using techniques like mineral fertilizers and pesticides. This negative definition has stuck until today, although organic production methods rely on the powerful positive concept of the links among fertile soils, healthy crops and livestock, healthy food and healthy human beings, a concept introduced by Albert Howard in the 1930s:

> [E]vidence for the view that a fertile soil means healthy crops, healthy animals, and healthy human beings is rapidly accumulating. At least half of the millions spent every year in trying to protect all three from disease in every form would be unnecessary the moment our soils are restored and our population is fed on the fresh produce of fertile land.
>
> (Howard, 1942, para 4)

This concept also included the idea of self-regulation in nature:

> The crops and livestock look after themselves. Nature has never found it necessary to design the equivalent of the spraying machine and the poison spray for the control of insect and fungus pests. There is nothing in the nature of vaccines and serums for the protection of the livestock. It is true that all kinds of diseases are to be found here and there among the plants and animals of the forest, but these never assume large proportions. The principle followed is that the plants and animals can very well protect themselves even when such things as parasites are to be found in their midst. Nature's rule in these matters is to live and let live.
>
> (Howard, 1940, pp. 3–4)

5.2 Problems of Early Organic Farms

5.2.1 Crop production and soil management

The early years of organic farming were characterized by striving to find self-regulation in crop production and soil management on many pioneer farms. However, its practical value was weak and some farmers failed, many with big financial losses. Managing a fertile soil was the guiding production strategy. To feed soil microorganisms with appropriate forms of organic matter, different combinations of manure, slurry and green mulch were prepared and tested. In order not to disturb the soil biota and mix soil layers with their characteristic physiological properties, sheet composting and shallow tillage were developed.

Although the idea of self-regulation was underpinned by many examples from farms, as well as what could be observed in nature, the reality was much more complex than the theory. The first problems occurred with weeds, which were difficult to deal with by the principle of 'live and let live', especially because what was effective against weeds (ploughing to bury their seeds and roots deeply) was bad for soil fertility. Around 30 years of research and development would be needed before ingenious weed prevention techniques became available, such as through adapted crop rotations and seedbed preparation, as well as implements for efficient and effective mechanical and thermal weed control.

An ideal organic farm in Europe was the mixed farm with a dairy cow herd fed on permanent grassland and clover–grass ley in the arable rotation. The main cash crops were wheat, barley, potatoes and some field vegetables (e.g. carrots, beets and various kinds of cabbages). Such a typical organic farm was also reflected in the design of the Haughley experiment, which was started in 1939 (Balfour, 1943), and which produced the most important data on the performance of organic farming until other comparative trials began in Sweden, Germany and Switzerland in the 1970s.

Very sensitive crops like grapes, apples, some berries and also most of the very challenging glasshouse vegetables (e.g. tomatoes, cucumbers) were marginal in organic farming at first. Their appearance was often so poor that they were not suitable for commercialization. Copper and sulphur fungicides, and occasionally the natural insecticides nicotine and rotenone, were used for controlling diseases and insects in these crops. Their ecological effects and their acute or chronic toxicity for human health were not considered. However, compared with the pesticides used in organic production, most of those used in conventional production were worse (e.g. organochlorine pesticides, which were banned and then replaced by organophosphates and carbamates).

5.2.2 Livestock production

Livestock husbandry was not a big or very controversial issue on early organic farms. Cattle were mainly seen as manure providers; therefore, the most important requirement for organic livestock was to feed them with farm-produced feeds to keep the nutrient cycles very local. Animal welfare in its holistic conception was not considered essential, health prevention strategies were not yet adopted and use of complementary medicine was marginal. In biodynamic agriculture, dairy cows were highly esteemed, especially because of their grazing and manuring activities, which were seen as the catalyst for fertility building in arable soils throughout human history. None the less, animal welfare was first emphasized by animal rights activists and by ethological researchers at universities, only by the 1960s and 1970s.

For both crops and livestock, organic farms were like islands where a new approach to soil and livestock husbandry could be learnt and demonstrated. Farms were preferably mixed and production enterprises that were difficult to manage passed over. Organic farming was not an industry, and the requirements of the food markets and the law of supply and demand did not apply. Therefore, it was not necessary to offer a complete assortment, continuous market supply, good shelf life, competitive prices and attractiveness to consumers (see Chapter 7, this volume, for a further description of the early organic market).

5.3 A Half-century of Progress

5.3.1 Crop production

In the 20th century, techniques such as the recycling of manure, management of natural soil fertility and use of cover crops tended to be neglected, as chemical and mineral fertilizers became widespread (Pound et al., 1999). The rejection of the cheap, but environmentally hazardous, fertilizers has made organic farms the practical 'laboratories' preserving and further developing knowledge of these techniques. In recent years, this knowledge of organic farmers, once ridiculed as 'old fashioned', has become a subject of fast-growing interest in crop science. It is valued for coping with environmental problems of conventional production or when making extensive forms of agriculture, especially in southern countries, more productive without increasing external inputs.

Farmyard manuring is sometimes regarded as synonymous with organic farming, although it is only one element of all soil fertility-building techniques. Manuring strategies are broadly described in the literature and advisory materials for farmers (see von Fragstein, 1995), and recent research work has primarily dealt with reducing nitrogen (N) losses from leaching and gaseous emissions (e.g. Philipps and Stopes, 1995). This research has improved the nutrient efficiency of the whole manure and slurry chain from livestock to plant uptake. The same goes for research on modern composting techniques based on the traditional knowledge of organic farmers and gardeners.

Equally important fertility management techniques are either well-designed crop rotations or catch crops and green manures. Hess (1990) studied N transfer along different organic crop rotations in temperate climatic zones in Germany. This and other work led to a better understanding of how to use the crop rotation to maximize yield and minimize nutrient losses. In their literature review, Thorup-Kristensen et al. (2003) use the term 'catch crop' for a cover crop grown to catch available N in the soil in order to prevent N leaching losses, and the term 'green manure' for a cover crop grown mainly to improve the nutrition of the succeeding main crop. The most common cover crops in central and northern Europe are crucifers (fodder radish, white mustard), monocots (ryegrasses, winter rye, oats) and legumes (hairy vetch, red, white, sweet and crimson clover, faba beans, field peas and lucerne). Advances in cover crops have made even stockless systems productive on organic farms (Schmidt *et al.*, 1999; Welsh *et al.*, 2002).

Since 1950, more than 500 papers relating to fertility-building crops have been published, 20% of them directly linked to organic farms or organically managed sites (Organic Soil Fertility, 2005). Plant nutrition in organic farming is an excellent example of how the practical knowledge set the research agenda and how science was able to catch up. Many aspects of sustainable soil fertility management have been either modelled or experimentally studied, such as biological N fixation, nutrients transferred between field crops, decomposition and mineralization processes in soil, nutrient recovery by plants and nutrient losses through leaching and volatilization (see Chapter 4, this volume, for some examples). A durable organic crop nutrition, emphasizing cycling within the farm system and maintaining a biologically active soil that can release nutrients from the soil, making them available for crop growth, has become feasible under very different climatic and site conditions (Mäder *et al.*, 2002).

Self-sustaining N supply in crops is a major innovation of organic farming and a step towards making agriculture independent of fossil energy supplies. In the near future, with rising oil prices and decreasing oil reserves, the organic technique could become a predominant one.

In contrast to N, phosphorus supply in sustainable systems has remained partly unsolved. In the small-scale farms of Europe, phosphorus supply is usually sufficient, thanks to rich soils, climatically favourable conditions and short cycles of phosphorus from livestock to crops in the form of manure (Fortune *et al.*, 2000). Depletion can be observed in the so-called broadacre organic farming in Australia (Penfold, 2000). It is also a major issue in many other countries, especially in the southern hemisphere. Research has shown the importance of manuring, soil organic matter content, plant root exudates and mycorrhizal fungi and other microorganisms, among others (Oberson *et al.*, 1993). However, the method to make the best use of these organic practices in inherently phosphorus-poor soils and under conditions not favourable for mixed livestock and crop production remains to be solved (Oberson *et al.*, 2000).

From the beginning, shallow ploughing has been the compromise adopted in organic farming to give good weed control by burying weed seeds and rhizomes while not disturbing soil animals and microorganisms too strongly by mixing soil layers with different physiological and ecological properties. Minimum or no tillage has not yet been successful in organic farming. These techniques work best in conventional systems, where herbicides and N fertilizer can be used to reduce weed competition and give a boost to otherwise low-yielding crops. Since minimum tillage is known to be more soil conserving than shallow ploughing, some farmers tried to modify it for organic farming. Several of these attempts were successful, and biodynamic farmers especially reported a positive interaction between no-tillage and biodynamic preparations. Lampkin (1990, p. 38) wrote that 'there is no reason why the concept should not be developed further within an organic context'. Although research activities have increased considerably since then, reduced or no-tillage techniques cannot be recommended for most organic farms. Depending on the crop, yield losses between 30% and 80% have been reported compared with a ploughed seedbed.

Ineffective weeding techniques had long made farmers very critical of organic farming, whether in arable farming, horticulture or permanent grassland. Conventional farming's dependence on hazardous chemicals was nowhere as obvious as with herbicides, which have brought about a huge saving in labour and enabled further industrialization of food production. But practical knowledge of how to manage weeds in organic farming by well-designed crop rotations, as well as by different mulching and tillage techniques, has improved considerably, and innovative implements for physical weed control have become so efficient that organic farms have become bigger too. Although in the 1980s, springtime harrows 6 m wide were common, the widths have doubled to 12 m since then, and the work rate has increased from 2 to 4 ha/h (Dierauer and Stöppler-Zimmer, 1994).

Novel techniques, such as the combination of harrowing and hoeing in cereals, and novel implements, such as brush or flame weeders, also have greatly improved weeding results in difficult crops such as vegetables, or in situations with high weed pressure. Consequently, weeds have become less threatening and are no longer a serious reason for farmers not to convert to organic farming. Bokhorst (1989) found that the labour needed per hectare of sugarbeet production decreased from 133 h in 1980 and 227 h in 1981 to 43 h in 1983 and 18 h in 1984. This reduction was due to an excellent weed prevention and management strategy (Fig. 5.1).

Driven by the farm machinery industry to quantify the progress scientifically, weed prevention and control in organic crops has been, and still is, a topic of research, mainly complementing work on innovations occurring on farms (HDRA Organic Weed Management, 2005). Perennial weeds, such as thistles in rotations dominated by wheat, or broad-leaved dock in leys and permanent grassland, have remained a serious problem so far. Besides economic problems, these weeds are the most frequent reasons that farmers have for leaving organic production.

Much less successful has been the self-regulation concept for controlling crop diseases and insect pests. The idea of 'fertile soil means

Fig. 5.1. Herbicides are banned on organic farms, but researchers and manufacturers have improved mechanical weeding techniques considerably. (Photo courtesy of FiBL. With permission.)

healthy plants' did not match the reality of daily experience on thousands of organic farms around the world. Howard had argued for a very holistic concept of plant health:

> Insects and diseases are not the real cause of plant diseases but only attack unsuitable varieties or crops imperfectly grown. Their true role is that of censors for pointing out the crops that are improperly nourished and so keeping our agriculture up to the mark. In other words, the pests must be looked upon as nature's professors of agriculture: as an integral portion of any rational system of farming... The policy of protecting crops from pests by means of sprays, powders, and so forth is unscientific and unsound as, even when successful, such procedure merely preserves the unfit and obscures the real problem – how to grow healthy crops.
> (Howard, 1940, p. 161)

A more diversified strategy has been developed, using habitat management or ecological engineering, prevention and sanitation techniques, and environmentally sound direct plant protection methods and agents. Habitat management or habitat manipulation represents a major approach of organic farmers to control pests (Fig. 5.2). Gurr *et al.* (2004) introduced the term 'ecological engineering' for that prevention strategy. They defined it as 'the design of sustainable systems consistent with

Fig. 5.2. Habitat management is an effective strategy against pests and diseases. (Photo courtesy of FiBL. With permission.)

ecological principles that integrate human society with its natural environment for the betterment of both'.

Tamm (2000) explained why many potentially noxious organisms are likely to cause much more serious problems in the future. The reasons are entirely man-made: higher specialization on the farm level; stricter market demands on the appearance of produce; the requirement of constant market supply; and changes in regulations on the international as well as national level. While the first three reasons are erosions of the principles of organic farming driven by the economy, the last reason represents a notable improvement in the application of organic principles: a ban on copper use in organic farming (still pending in the European Union (EU) at the moment) and the requirement that seeds, seedlings and propagation materials be organically produced (and therefore not treated with fungicides) will give rise to genuine organic solutions and treatments.

Tamm (2000) outlined a modern concept involving a combination of the following elements and actions:

- **Sanitation**, which includes strategies such as: use of high-quality seeds (e.g. against the fungus causing common bunt, *Tilletia tritici*, in wheat); removal of overwintering sources of inoculum (e.g. the brown rot fungus, *Monilia laxa*, in sweet cherry); and removal of

infected volunteer plants (e.g. the late blight fungus, *Phytophthora infestans*, in potato).
- **Avoidance techniques**, which are usually achieved by exposing a crop at later physiological stages to a noxious organism (e.g. chitting of seed potato).
- **Tolerant or resistant varieties**, which remain the backbone of organic agriculture, provided that the available varieties with resistance (e.g. against the apple scab fungus, *Venturia inaequalis*) are acceptable to growers and consumers.
- **Variety mixtures**, which has become a very efficient technique to stabilize the yields and quality of certain crops; in Switzerland, for instance, more than 90% of organic wheat is grown in variety mixtures.
- **Intercropping**, a technique that has not yet been applied widely; whereas the control of fungal diseases by this strategy may be difficult, entomologists have achieved spectacular successes in suppression of certain insect pests (e.g. the rosy apple aphid, *Dysaphis plantaginea*, in apple).
- **Soil management and plant nutrition**, whose impact is well known in principle, but which still have a huge potential to be exploited (e.g. against *Phytophthora infestans* in tomato).
- **Crop protection agents**, such as fungicides, insecticides, antagonists or inducers of resistance, which are the only remaining solution if none of the strategies mentioned above leads to acceptable control. For instance, the development of a biocontrol agent and introduction of a neem-based product against the codling moth (*Cydia pomonella*) and aphids, have facilitated reliable organic production of apples in Switzerland.

'Healthy plants' still offer scientists many problems to resolve. Therefore, several research priorities of the 5th and 6th Framework (2000–2008) of the European Commission emphasize plant protection in organic systems, especially alternatives to copper fungicides in potatoes and grapes, and scientific criteria for the evaluation of plant protection products used in organic agriculture.

The rapid growth of organic production and the more impersonal food supply chain of conventional supermarkets with organic programmes have accentuated the economic relevance of insufficient disease control. Blemishes, spots, maggots or adult insects are not tolerated by consumers remote from direct information about the farmers.

Various scientific studies have shown how significant a fertile soil and proper crop management are in making crops more tolerant to diseases. Molecular and physiological studies have explained in detail how plants reactivate their immune systems when attacked by a fungus, bacteria, nematode or insect. Some defence mechanisms of plants are passive, while others are induced when the plant detects a pathogenic agent. One strategy for crop protection, which also takes into account environmental considerations, is to mimic a pathogen attack in order to trigger defence responses before possible infection. Elicitors are molecules that are capable of doing this by mimicking the detection of a pathogen by the plant (Dangl and Jones, 2001).

Many pest control strategies in organic farming are thought to use these mechanisms, because soil microorganisms (common in a fertile soil) can also function as elicitors of plant resistance. Also, there are natural compounds that can be applied on leaves to trigger the defence reaction of plants, e.g. salicylic acid from willow trees or water extracts from *Penicillium mycelium* (Rentsch, 1998).

The knowledge of systemic acquired resistance in plants is a product of cutting-edge science and reflects Howard's concept of a healthy plant. Therefore, it is not very surprising that long-term organic farmers insist that the longer a farm is managed organically and the better the management respects soil fertility principles, the less threatening the disease problems become. This intuitive feeling was substantiated by long-term experiments and by research on pioneer farms (Berner *et al.*, 2003). The principle that 'fertile soils mean healthy plants' seems to distinguish long-term organic farms from newly converted ones.

5.3.2 Livestock production

Innovation in livestock husbandry did not initially come from the organic pioneers, although respect towards animals and the idea of some kind of co-evolution of humans and livestock have been themes of organic farming, especially biodynamic agriculture, from the very beginning. Industrialized and mechanized livestock production evoked public outrage starting in the 1970s in many European countries, the USA and Canada. Many non-governmental organizations (NGOs) became involved either as animal rights activists or as promoters of animal welfare or free-range production systems, which led to labelling and marketing activities. Free-range, 'happy' laying hens appeared in many European countries long before organic eggs were commercialized. This drove universities to study more intensively the behaviour of domestic farm animals. Ethological research quickly delivered a practical framework for animal welfare. In northern Europe, where tethering of dairy cows was traditional, even organic farmers initially felt that they were being pushed too hard with animal welfare concepts. But the opposition did not last long, and animal welfare became a distinctive mark of organic farms, leading to constant amendment of national, EU and international standards (IFOAM and Codex Alimentarius, the food and veterinary standards of FAO and WHO), especially between 1990 and 2000 (see Chapter 8, this volume).

The changes in organic livestock husbandry from traditional tethering systems towards ethologically appropriate housing and free-range systems, primarily driven by consumer concern, substantiated by research activities and guided by the standardization process, is one important area where the progress did not originate primarily from the bottom up. Hence, recent discussions of animal welfare concepts for organic farms in Scandinavian countries surprisingly reveal new ideas developed jointly by farmers and researchers. Alrøe et al. (2001) saw a strong link between the organic concept of 'naturalness' (which was used mainly for crops, food quality and processing) and the possibilities offered to livestock to express their natural behaviour, including natural reproduction and growth. From an organic perspective, they added additional criteria like 'harmony' and 'care'. Harmony mirrors the interactions between the farm and its environment, among the different elements of the farm, and among the animals in the herd. Care addresses the special responsibility that farmers, transporters and processors have towards domestic animals.

Compared with animal welfare, health concepts for organic herds are less developed and are more challenging to implement. Because diseases cause livestock to suffer and become weak, interventions with conventional allopathic drugs such as antibiotics are not totally banned in most standards, including those of the EU. Rather, they are restricted in how frequently they may be used, and their withholding periods are doubled. Good livestock health does not simply mean the absence of disease, but also a high level of vigour and vitality, thus enhancing the animal's ability to resist infections, parasitic attacks and metabolic disorders, and to recover from injury. However, the real situation in organic herds in most regions of the world does not yet reflect this idea. Even with the National Organic Standards in the USA, which ban antibiotic use absolutely, this has not yet led to a holistic health strategy, as infected animals are simply removed from organic herds rather than cured.

Younie (2000) summarized the elements of a preventive health strategy as follows:

- Self-contained herds and flocks
- Appropriate choice of breed
- Breeding for disease and parasite resistance
- Suckling with mother
- Natural weaning
- Access to pasture during the growing season

- Adequate nutrition: high forage, limited cereals
- Regular monitoring of feed, physiological status and health (e.g. silage, milk and urine; faecal worm egg counts)
- Establishment of a clean grazing system: low stocking rates, alternating from year to year (sheep/cattle), mixed grazing (sheep/cattle), mixed age groups, use of hay/silage aftermaths
- Adequate space, good ventilation and adequate supply of dry bedding if indoor accommodation is required

While preventive strategies have been successfully established in arable crops and later in horticultural crops, organic farmers still face severe problems when applying holistic concepts of livestock health. Much the same range of diseases and parasite problems occurs in organic as in conventional livestock. Consequently, the use of antibiotics and anthelmintics has not yet been reduced to zero or replaced by organically appropriate materials and treatments. These two groups of drugs – antibiotics targeted mainly towards udder diseases of dairy cows, sheep and goats, and anthelmintics, targeted mainly towards internal parasites of sheep, cattle, pigs and poultry – are the predominant forms of therapy used on organic farms. Several surveys in European countries that are representative of the health status as well as the kind and frequency of allopathic medications worldwide revealed no substantial differences between organic and conventional herds, e.g. in the UK (Halliday et al., 1991; Roderick et al., 1996), Germany (Krutzinna et al., 1996; Brinkmann and Winckler, 2005), the Netherlands (Kijlstra et al., 2003), Austria (Baumgartner et al., 2001), Switzerland (Busato et al., 2000), Norway (Hardeng and Edge, 2001), Denmark (Bennedsgaard et al., 2003) and Sweden (Höglund et al., 2001).

Keatinge et al. (2000, p. 95) summarized the health status on organic farms as follows: 'Taken as a whole, the information indicates that within current standards the incidence of disease on organic farms is generally at acceptable levels, and at least in some specific circumstances, better than that on conventional farms. The data also indicate a significant use of allopathic medicine.'

To achieve holistic and systemic health management, preventive measures and alternative therapy forms have to be combined. Alternative therapy is still in its infancy. Clinical studies on homoeopathic treatments of farm animals give a very inconsistent picture. A critical meta-study of all clinical studies is still missing. The only meta-study known to the author is a PhD thesis (Kowalski, 1989) that is not available and was never published. Many clinical studies have methodological problems and few are completely standardized (Fidelak et al., 2003).

Organic farmers have remarkably good traditional knowledge on how to cope with endoparasites of cattle, especially gastrointestinal worms and lungworms. The crucial element of their intuitive but very successful prevention is the stocking density, as several in-depth studies have shown (Thamsborg and Roepstorff, 2003; Hördegen, 2005). A low stocking density and consequently a lower infection rate can be obtained by alternating grazing and mowing, by mixing species (cattle and sheep) or by mixing different age groups of the same species with different susceptibility to infection (e.g. yearling heifers and second-year heifers). However, these management measures are not sufficient in the case of liver fluke.

Major problems also remain with sheep, pigs and poultry, where preventive measures have to be combined with direct control. In contrast to mastitis, it seems to be easier to develop organically appropriate agents against worms (Thamsborg and Roepstorff, 2003). It can be expected that the different research groups working on this economically important group of parasites will soon come up with a combination of preventive measures and natural wormers, such as plant extracts (Hördegen *et al.*, 2003) and living organisms. Whereas plant extracts will be applied by way of bioactive forages (e.g. plants with a high content of condensed tannins such as *Lotus* spp. or *Sulla*), living organisms such as the nematode-trapping fungus *Duddingtonia flagrans* are added to the forage.

Livestock health is a critical problem for the protection of animals. Pests and diseases cause pain and therefore demand an adequate and fast cure. Unlike with crop production, the organic pioneers did not put the same effort into developing effective preventive strategies in livestock husbandry. This leads to a divergence between what is claimed to be an essential feature of organic farming (no use of synthetic chemical inputs) on one hand and the reality of organic livestock husbandry on the other. Organic dairy farmers in the USA have adopted a throwaway approach, either culling sick cows or moving them to conventional farms. Most European producers prefer a more sustainable health management concept, but this still has to be developed. As mastitis, like many other diseases, is a multifactorial problem, interdisciplinary research at its best is needed, bringing together veterinarians, agronomists and economists. It also needs an approach that takes into account that farmers and practical veterinarians are the main actors in a successful health strategy (Vaarst *et al.*, 2004). Researchers still have a long way to go, and it is unfortunate that such research started only in the 1990s.

5.4 Issues in Organic Research

Genetically engineered organisms have partially succeeded pesticides and N fertilizers as the scapegoats of modern intensive agriculture.

In the 1970s, when the negative impacts of pesticides and fertilizers on the environment and on natural and semi-natural habitats became obvious, the chemical companies developed the fascinating plan of incorporating protection of the plant within the plant itself by genetically engineering it to produce pesticides, such as the *Bacillus thuringiensis* (Bt) toxin. To some extent this idea also applied to the plant's nutrition, such as by creating N-fixing maize and other non-legumes. This seemed to promise an end to releasing some hazardous substances into the environment.

In the first round, individual representatives of the organic movement were not strictly opposed to the molecular approach, as it seemed to offer many pesticide-free solutions to agronomic problems, not only in conventional but also in organic production. It took a while before the emerging technology was understood in its complex and irreparable impacts on ecosystems and along the whole food chain, and therefore was banned in organic farming. As in its earliest days, organic farming erroneously reinforced its negative image, which often makes it difficult to communicate its innovative and challenging concepts of sustainability. Organic farming is not a matter of saying no to technology, but rather a comprehensive agroecological strategy for agriculture in general.

To foster the positive messages of organic farming, a major challenge for both research and practical farming will be the 'ecological engineering' approach to system design (Gurr *et al.*, 2004; Zehnder *et al.*, 2007), as described earlier in this chapter. Coping with agronomic problems through systemic stability will change organic farms. Interactions among all parts of the farm and its natural and semi-natural environment are organized along complex networks, not linearly. Geographical Information Systems (GIS) and different tools of modern biostatistics are potentially useful to speed up system research, but are not fully used by many research groups. Organic crop and animal husbandry techniques focusing on prevention instead of interventions are exactly what consumers increasingly demand of foods in general. Pesticide-free production and additive-free processing are trends that also influence organic food and farming. It is difficult and extremely expensive to develop organic agents for crop and livestock production such as plant extracts, products from microorganisms, and living organisms, because sales of such products are limited to a niche and therefore are not profitable. In addition, consumers' scepticism caused by confusing lists of hundreds of chemical compounds used during the life cycle of conventional foods will also lead to rejection of natural and organic inputs.

Breeding will become a major approach in organic farming to overcome technical obstacles in organic crop and livestock husbandry and make the farms less dependent on external inputs. Biodynamic plant breeders started essential work in the late 1980s, with Germany, Switzerland and the Netherlands becoming focal points (Röckl, 2002). On-farm selection plays an important role among biodynamic breeders. Farmers' knowledge and practical experience regarding the performance, adaptability to management and site conditions, and the genuine resilience of breeds and lines are intensively used by the breeders. Therefore, biodynamic plant breeding represents participatory research in its purest and best form, and has had tangible success. For example, the Swiss biodynamic breeder Peter Kunz introduced wheat and spelt cultivars that performed much better on low-input farms in yield, stability and quality than the best conventional breeds (Getreidezüchtung Peter Kunz, 2005). This success was recently topped by a wheat cultivar resistant to common bunt caused by *Tilletia tritici*, whereas traditional plant breeders still hope to solve the problem using genetic engineering, expressing KP4 proteins from strains of *Ustilago* in wheat.

Moreover, in livestock breeding, the dominance of productivity measures in breeding programmes has been questioned recently. In several countries, such as Austria, Germany, the Netherlands and Switzerland, the debate was initiated by farmers. Additional traits like health status, the ability to recover from management and environmental stress, as well as excellent adaptability to a low-input environment and organic feeding practices are being discussed by farmers, researchers and other experts (Hovi and Baars, 2001). These first timid steps aim at better embedding livestock into the specific environment offered by organic farms, which is characterized by free-range conditions, more spacious housing, less concentrated feeds and completely different strategies for disease prevention and therapy.

Biodynamic farmers are the driving force behind the changes in breeding priorities and techniques for both crops and livestock, at least in Europe. In contrast to purists seeking to conserve old varieties, these farmers try to move breeding towards low-input production systems and high-output/high-quality cultivars. Among scientists who try to accommodate organic breeding strategies, both breeding aims and techniques are the subject of controversy. Is it a more holistic approach to use field selection and emphasize phenotype features and genotype–environment interactions, or does the use of DNA markers speed up the breeding process towards organically desirable traits? By using a combination of both approaches, organic farming will become truly innovative.

5.5 A Future Challenge for Organic Farming Research

Seen in historical perspective, organic farming practice and science have evolved surprisingly smoothly and in harmony. Cooperation has been characterized by partnership and utmost solidarity, which might be because both farmers and scientists were once outsiders in their respective professional communities. In such a 'community of fate', cooperation was taken for granted. Yet no original theory and practice of participatory research has been developed by organic farming to the degree that it has in nature conservation and landscape research, for example, or for international research activities in the tropics and subtropics (Gottret and White, 2001; Gonsalves et al., 2005).

That is one reason Lockeretz (2000) quite rightly concluded that the main difference between organic and conventional farming research is in what gets studied, not how one studies it. As the scientific community of organic researchers grows steadily and becomes more a part of the mainstream, an alienation will occur between those who produce theories and research results and those who use them. This development will be speeded up by the fact that organic scientists, once they are outside their original niche, have to follow the rules of the wider community. Interdisciplinary or even participatory research is not the priority of prestigious scientific journals where one's work will be frequently cited. For the future success of organic farming, it will be important to maintain and even intensify the fruitful exchange of knowledge between farmers and scientists (Alrøe and Kristensen, 2002; Baars, 2002).

References

Alrøe, H.F. and Kristensen, E.S. (2002) Towards a systemic research methodology in agriculture: rethinking the role of values in science. *Agriculture and Human Values* 19, 3–23.

Alrøe, H.F., Vaarst, M. and Kristensen, E.S. (2001) Does organic farming face distinctive livestock welfare issues? A conceptual analysis. *Journal of Agricultural and Environmental Ethics* 14, 275–299.

Baars, T. (2002) Reconciling scientific approaches for organic farming research. Part I: Reflection on research methods in organic grassland and animal production at the Louis Bolk Institute, The Netherlands. PhD thesis. Publication Nr. G38, Louis Bolk Instituut, Driebergen, The Netherlands.

Balfour, E.B. (1943) *The Living Soil*. Faber & Faber, London.

Baumgartner, J., Leeb, T., Guber, T. and Tiefenbacher, R. (2001) Pig health and health planning in organic herds in Austria. In: Hovi, M. and Vaarst, M. (eds) *Positive Health: Preventive Measures and Alternative Strategies*. Proceedings of the Fifth NAHWOA Workshop, Rødding, Denmark, 11–13 November 2001. University of Reading, Reading, UK, pp. 126–131.

Bennedsgaard, T.W., Thamsborg, S.M., Aarestrup, F.M., Enevoldsen, C., Vaarst, M. and Larsen, P.B. (2003) Use of veterinary drugs in organic and conventional dairy herds in Denmark with emphasis on mastitis treatment. Available at: orgprints.org/4712/

Berner, A., Gloor, S., Fuchs, J.G., Tamm, L. and Mäder, P. (2003) Gesunder Boden – gesunde Pflanzen. In: Freyer, B. (ed.) *Beiträge zur 7. Wissenschaftstagung zum ökologischen Landbau: Ökologischer Landbau der Zukunft. Wien, 24–26 Februar 2003.* Institut für Ökologischen Landbau, Universität für Bodenkultur, Vienna, pp. 443–444. Available at: orgprints.org/2796/02/berner-et-al-2003-gesunder-boden-pflanze-en.pdf

Bokhorst, J.G. (1989) The organic farm at Nagele. In: Zadocks, J.C. (ed.) *Development of Organic Farming Systems: Evaluation of the Five-year Period 1980–1984.* Pudoc, Wageningen, The Netherlands, pp. 57–65.

Brinkmann, J. and Winckler, C. (2005) Status quo der Tiergesundheitssituation in der ökologischen Milchviehhaltung – Mastitis, Lahmheiten, Stoffwechselstörungen. In: Hess, J. and Rahmann, G. (eds) *Ende der Nische. Beiträge zur 8. Wissenschaftstagung Ökologischer Landbau.* Kassel University Press GmbH, Kassel, Germany, pp. 343–346.

Busato, A., Trachsel, P., Schällibaum, M. and Blum, J.W. (2000) Udder health and risk factors for subclinical mastitis in organic dairy farms in Switzerland. *Preventive Veterinary Medicine* 44, 205–220.

Dangl, J.L. and Jones, J.D.G. (2001) Plant pathogens and integrated defence responses to infection. *Nature* 411, 826–833.

Dierauer, H.-U. and Stöppler-Zimmer, H. (1994) *Unkrautregulierung ohne Chemie.* Ulmer, Stuttgart, Germany.

Fidelak, Chr., Paal, K., Merck, C.C., Klocke, P. and Spranger, J. (2003) Homoeopathy in bovine mastitis – a randomised placebo controlled double-blind study. In: *Improving the Success of Homoeopathy* 4. The Royal London Homoeopathic Hospital, London, pp. 97–98.

Fortune, S., Conway, J.S., Philipps, L., Robinson, J.S., Stockdale, E.A. and Watson, C. (2000) N, P and K for some UK organic farming systems – implications for sustainability. In: Rees, R.M., Ball, B., Watson, C. and Campbell, C. (eds) *Soil Organic Matter and Sustainability.* CAB International, Wallingford, UK, pp. 286–293.

Getreidezüchtung Peter Kunz (2005) Available at: www.peter-kunz.ch

Gonsalves, J., Becker, Th., Braun, A., Campilan, D., de Chavez, H., Fajber, E., Kapiriri, M., Rivaca-Caminade, J. and Vernooy, R. (2005) *Participatory Research and Development for Sustainable Agriculture and Natural Resource Management: A Sourcebook. Volume 1: Understanding Participatory Research and Development.* International Potato Centre – Users' Perspectives with Agricultural Research and Development, Laguna, Philippines, and International Development Research Centre, Ottawa, Canada.

Gottret, M.A.V.N. and White, D. (2001) Assessing the impact of integrated natural resource management: challenges and experiences. *Conservation Ecology* 5, 17. Available at: www.consecol.org/vol5/iss2/art17/

Gurr, G.M., Wratten, S.D. and Altieri, M.A. (2004) *Ecological Engineering for Pest Management. Advances in Habitat Manipulation for Arthropods.* CAB International, Wallingford, UK.

Halliday, G., Ramsay, D.A., Scanlan, S. and Younie, D. (1991) *A Survey of Organic Livestock Health and Treatment.* Kintail Land Research Foundation, Glasgow, UK, in association with Scottish Agricultural College.

Hardeng, F. and Edge, V.L. (2001) Mastitis, ketosis, and milk fever in 31 organic and 93 conventional Norwegian dairy herds. *Journal of Dairy Science* 84, 2673–2679.

HDRA Organic Weed Management (2005) Available at: hdra.org.uk/organicweeds/

Hess, J. (1990) Acker- und pflanzenbauliche Strategien zum verlustfreien Stickstofftransfer beim Anbau von Kleegras im Organischen Landbau. *Mitteilungen Gesellschaft Pflanzenbauwissenschaften* 3, 241–244.

Höglund, J., Svensson, C. and Hessle, A. (2001) A field survey on the status of internal parasites in calves on organic farms in Southwestern Sweden. *Veterinary Parasitology* 99, 113–128.

Hördegen, P. (2005) Epidemiology of internal parasites on Swiss organic dairy farms and phytotherapy as a possible worm control strategy. PhD thesis No. 16144, ETH Zürich, Switzerland.

Hördegen, P., Hertzberg, H., Heilmann, J., Langhans, W. and Maurer, V. (2003) The anthelmintic efficacy of five plant products against gastrointestinal trichostrongylids in artificially infected lambs. *Veterinary Parasitology* 117, 51–60.

Hovi, M. and Baars, T. (eds) (2001) *Breeding and Feeding for Animal Health and Welfare in Organic Livestock Systems*. Proceedings of the Fourth NAHWOA Workshop, Wageningen, 24–27 March 2001. The University of Reading, Reading, UK.

Howard, A. (1940) *An Agricultural Testament*. Oxford University Press, London. Reprinted 1972 by Rodale Press, Emmaus, Pennsylvania.

Howard, A. (1942) Medical testament. In: Grant, D. (ed.) *Feeding the Family in War-time, Based on the New Knowledge of Nutrition*.

Keatinge, R., Gray, D., Thamsborg, S.M., Martini, A. and Plate, P. (2000) EU Regulation 1804/1999: the implications of limiting allopathic treatment. In: Hovi, M. and Garcia Trujillo, R. (eds) *Diversity of Livestock Systems and Definition of Animal Welfare*. Proceedings of the Second NAHWOA Workshop, Cordoba, 8–11 January 2000. University of Reading, Reading, UK, pp. 92–98.

Kijlstra, A., Groot, M., van der Roest, J., Kasteel, D. and Eijck, I. (2003) Analysis of black holes in our knowledge concerning animal health in the organic food production chain. Report Animal Sciences Group, Wageningen UR, The Netherlands. Available at: http://orgprints.org/1034/

Kowalski, M. (1989) Homöopatische Arzneimittelanwendung in der veterinärmedizinischen Literatur. PhD thesis, Freie Universität Berlin.

Krutzinna, C., Boehncke, E. and Herrmann, H.-J. (1996) Organic milk production in Germany. *Biological Agriculture and Horticulture* 13, 351–358.

Lampkin, N. (1990) *Organic Farming*. Farming Press Books, Ipswich, UK.

Lockeretz, W. (2000) Organic farming research, today and tomorrow. In: Alföldi, T., Lockeretz, W. and Niggli, U. (eds) *The World Grows Organic*. Proceedings of the 13th IFOAM International Scientific Conference. Basel, 28–31 August 2000. vdf Hochschulverlag AG an der ETH Zürich, Switzerland, pp. 718–720.

Mäder, P., Fließbach, A., Dubois, D., Gunst, L., Fried, P. and Niggli, U. (2002) Soil fertility and biodiversity in organic farming. *Science* 296, 1694–1697.

Oberson, A., Fardeau, J.C., Besson, J.-M. and Sticher, H. (1993) Soil phosphorus dynamics in cropping systems managed according to conventional and biological agricultural methods. *Biology and Fertility of Soils* 16, 111–117.

Oberson, A., Oehl, F., Langmeier, M., Fließbach, A., Dubois, D., Mäder, P., Besson, J.-M. and Frossard, E. (2000) Can increased soil microbial activity help to sustain phosphorus availability? In: Alföldi, T., Lockeretz, W. and Niggli, U. (eds) *The World Grows Organic*. Proceedings of the 13th IFOAM International Scientific Conference. Basel, 28–31 August 2000. vdf Hochschulverlag AG an der ETH Zürich, Switzerland, pp. 27–28.

Organic Soil Fertility (2005) Available at: www.organicsoilfertility.co.uk/home/index.html

Penfold, Chr. (2000) *Phosphorus Management in Broadacre Organic Farming Systems*. A report for the Rural Industries Research and Development Corporation, Australian Government, Kingston ACT 2604. Available at: www.rirdc.gov.au/comp00/org1.html_Ref 487529629

Philipps, L. and Stopes, C. (1995) The impact of rotational practice on nitrate leaching losses in organic farming systems in the United Kingdom. *Biological Agriculture and Horticulture* 11, 123–134.

Pound, B., Anderson, S. and Gundel, S. (1999) Species for niches: When and for whom are cover crops appropriate? *Mountain Research and Development* 19, 307–312.

Rentsch, C. (1998) Induced resistance caused by PEN in tomato (*Phytophthora infestans*), cucumber (*Colletotrichum lagenarium/Pseudoperonospora cubensis/Erysiphe* sp.) and grapevine (*Plasmopara viticola*). Diploma work, Basel University, Basel, Switzerland.

Röckl, C. (2002) Is organic plant breeding a public affair? In: Lammerts van Bueren, E.T. and Wilbois, K.P. (eds) *Organic Seed Production and Plant Breeding – Strategies, Problems and Perspectives*. Proceedings of ECO-PB First International Symposium on Organic Seed Production and Plant Breeding, Berlin, Germany, 21–22 November 2002. European Consortium for Organic Plant Breeding, Driebergen/Frankfurt, pp. 50–54. Available at: www.eco-pb.org

Roderick, S., Short, N. and Hovi, M. (1996) Organic livestock production: animal health and welfare research priorities. Veterinary Epidemiology and Economics Research Unit, Department of Agriculture, University of Reading, Reading, UK.

Schmidt, H., Philipps, L., Welsh, J.P. and von Fragstein, P. (1999) Legume breaks in stockless organic farming rotations: nitrogen accumulation and influence on the following crops. *Biological Agriculture and Horticulture* 17, 159–170.

Tamm, L. (2000) The impact of pests and diseases in organic agriculture. In: British Crop Protection Council (eds) *Pests and Diseases 2000*. Conference Proceedings, Vol. 1, Brighton, UK, pp. 159–166.

Thamsborg, S.M. and Roepstorff, A. (2003) Parasite problems in organic livestock and options for control. *Journal of Parasitology* 89(Suppl.), 277–284.

Thorup-Kristensen, K., Magid, J. and Jensen, L.S. (2003) Catch crops and green manures as biological tools in nitrogen management in temperate zones. *Advances in Agronomy* 79, 227–302.

Vaarst, M., Wemelsfelder, F., Seabrook, M., Boivin, X. and Idel, A. (2004) The role of humans in the management of organic herds. In: Vaarst, M., Roderick, S., Lund, V. and Lockeretz, W. (eds) *Animal Health and Welfare in Organic Agriculture*. CAB International, Wallingford, UK, pp. 205–225.

von Fragstein, P. (1995) Manuring, manuring strategies, catch crops and N-fixation. *Biological Agriculture and Horticulture* 11, 275–287.

Welsh, J.P., Philipps, L. and Cormack, W.F. (2002) The long-term agronomic performance of organic stockless rotations. In: Powell, J. (ed.) *Proceedings of the UK Organic Research 2002 Conference, Research in Context, 26–28 March 2002, Aberystwyth*. Organic Centre Wales, Aberystwyth, UK, pp. 47–50.

Younie, D. (2000) Integration of livestock into organic farming systems: health and welfare problems. In: Hovi, M. and Garcia Trujillo, R. (eds) *Diversity of Livestock Systems and Definition of Animal Welfare*. Proceedings of the Second NAHWOA Workshop, Cordoba, 8–11 January 2000. University of Reading, Reading, UK, pp. 13–21.

Zehnder, G., Gurr, G.M., Kühne, S., Wade, M.R., Wratten, S.D. and Wyss, E. (2007) Arthropod pest management in organic crops. *Annual Review of Entomology* 52, 57–80.

6 The Development of Governmental Support for Organic Farming in Europe

S. Padel[1] and N. Lampkin[2]

[1]*Research Associate, Organic Research Group, Institute of Rural Sciences, University of Wales, Aberystwyth SY23 3AL, UK;* [2]*Senior Lecturer in Agricultural Economics and Director, Organic Research Group, Institute of Rural Sciences, University of Wales, Aberystwyth SY23 3AL, UK*

6.1 The Growth of Organic Farming in Europe

Although organic farming has existed as a concept for over 70 years, only since the mid-1980s has it become the focus of significant attention from European policy makers, consumers, environmentalists and farmers. The European organic sector has grown significantly since the 1980s, which can partially be attributed to the growing recognition by policy makers (Dabbert et al., 2004; Dimitri and Oberholzer, 2005). In 1985, certified and policy-supported organic production accounted for only 105,000 ha in the European Union (EU) plus Norway and Switzerland, or less than 0.1% of the total agricultural area. By the end of 2004, this had increased to 6.2 million ha, nearly 4% of Utilizable Agricultural Area (UAA). During the same period, the number of organic holdings increased from 6600 to 164,000.

These figures do not reveal the great variability within and between countries. In 2004, 6–12% of UAA was managed organically in several countries (see Fig. 6.1), and more than 30% in some regions (Olmos and Lampkin, 2005).

Alongside these increases in the supply base, the market for organic produce has also grown significantly. Accurate statistics on the overall size of the market for organic produce in Europe remain rare (Hamm and Gronefeld, 2004), but it was estimated at €12–12.5 billion in 2004, second only to the US market (Willer and Yussefi, 2006), compared with less than €900 million in the early 1990s.

The dramatic growth in the organic sector since the early 1990s in Europe is related to a combination of policy support, growing consumer

©CAB International 2007. *Organic Farming: an International History* (Lockeretz)

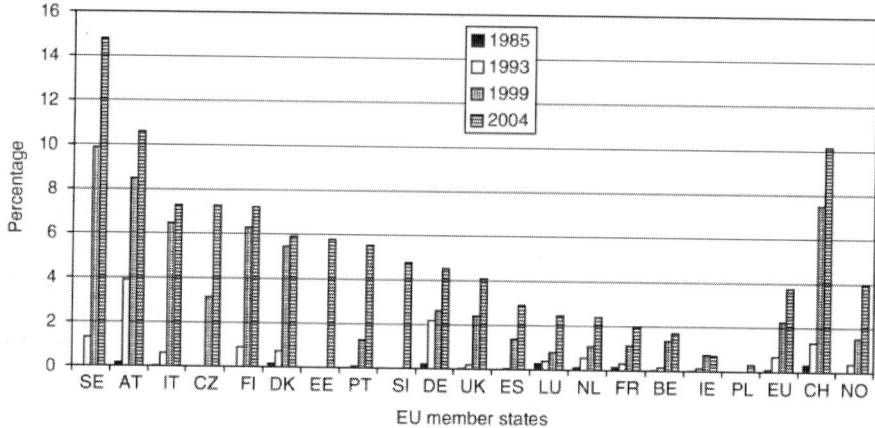

Fig. 6.1. Organic and in-conversion land area as percentage of utilized agricultural area (UAA) in selected European countries. (From Praznan and Koutna, 2004; Olmos and Lampkin, 2005.)

demand, and animal health, food safety and economic problems in the conventional agricultural sector. Some European countries have supported farmers in converting to or continuing with organic farming since the late 1980s, combined with some other policies aimed at promoting the organic sector. Because organic farming policy support originated mainly in Europe, we have focused this chapter on the European experience, although countries in other regions have also introduced support programmes in recent years (OECD, 2003; Dabbert et al., 2004).

Policy support for organic farming takes several forms, including legal standards defining organic farming; government inspection, certification and labelling activities; and direct payments to converting and established organic producers. Organic farming is also affected by the mainstream measures of the Common Agricultural Policy (CAP) and some special provisions introduced to mitigate these effects, as well as by the use of rural development programmes to support the development of marketing infrastructure and organic institutional capacity building, by information activities and research and by the balancing of various measures in integrated action plans.

6.2 Reasons for Policy Support for Organic Production

In the post-1945 period, most European governments were initially indifferent or even hostile towards organic farming, and interpreted this social movement as threatening the mainstream agricultural policy that emphasized technological progress as a means of improving farm incomes and securing food supplies. However, with the emergence of significant overproduction and environmental problems as a result of intensification, organic farming gained credibility as a possible alternative development path, particularly in the German-speaking and Nordic countries.

In recognition of the growing consumer interest in organic food and organic farming's potential contribution to policy objectives regarding surplus reduction, the environment and rural development, some European governments started the first policy initiatives for national regulations defining organic production in the mid-1980s, followed soon after by national programmes to aid conversion. In the early 1990s, EU-wide regulations were introduced defining organic crop production and providing support for converting to and continuing with organic production. By the turn of the millennium, the organic movement had moved from being seen as in opposition to mainstream agriculture policy to itself being the subject of significant policy intervention. This has led to much discussion of whether the movement can maintain control over its destiny, whether its principles have been eroded, whether farmers are still converting for the 'right' reasons and whether policy has gained too much influence over the organic sector (Dabbert et al., 2004).

Apart from the relatively short-lived idea that the lower yields from organic farming might help reduce the overproduction problem, European policy makers became interested in supporting organic agriculture for two main reasons (Dabbert et al., 2001). First, it was seen as a public good, delivering environmental, social and other benefits to society that at most are only partially paid for through the normal price of food. Second, it was an infant industry, support for which can be justified in terms of expanding consumers' choices and allowing the industry to develop to a point at which it could independently compete in established markets and make a positive contribution to rural development.

Although both justifications have been used in most countries, the first is more typical of some Scandinavian and Central European countries (e.g. Sweden, Finland, Austria), while the second approach is reflected in the Dutch focus on supply chain initiatives (MLNV, 2000) and the UK's initial unwillingness to support farms beyond the initial conversion phase (Lampkin et al., 1999a).

These main justifications for supporting organic farming are linked to the general issue of market failure, although unlike other agri-environmental policy measures, organic farming has developed a strong reliance on markets' and consumers' willingness to pay in support of its broader health, food quality and sustainability objectives (for a more detailed discussion of research into each of the separate arguments for policy support see Padel et al., 2002a; Dabbert et al., 2004).

It can be argued that the market-led strategy has been so successful that there may be significant risks associated with the market for organic products becoming an end in itself, rather than a means to achieve broader goals of benefit to society as a whole. The challenge for policy makers has been to develop a mix of policies that can use the market effectively, while allowing organic agriculture to remain true to its original aims, thus maximizing its broader benefits to society.

6.3 Legislation Defining Organic Production at the National and European Level

6.3.1 Providing a legal foundation for organic standards

In most European countries, private sector bodies were important in developing organic standards (see Chapter 8, this volume). However, because of growing consumer interest in organic products and because of the need to define which farmers would be eligible for conversion support programmes (see Section 6.4.1), the governments of five EU member states (Austria, Denmark, Spain, Finland and France) and Switzerland recognized a need to anchor national standards in law, and introduced national, legally enforceable definitions of organic production and in some cases also national certification procedures and labels.

In France, organic farming was first recognized in the policy context in 1980 in the Agricultural Law No 80502 (4 July 1980). The national logo for organic products (*Agriculture Biologique*) was introduced in 1984; detailed standards for crops were officially approved in 1986; standards for livestock production followed in 1992 (Lampkin *et al.*, 1999a). In Austria, the initiative to develop a regulation came from the Ministry of Consumer Protection, which created an organic section in the national food law, the Austrian Codex Alimentarius, in 1983 (Michelsen *et al.*, 2001). The 1987 Danish law on organic farming gave guarantees to consumers, defined the principles of organic farming and introduced public certification. It also set out the conditions that producers receiving payments for conversion to organic production had to follow. In 1990, before the EU regulation of 1991, Spain and Finland also introduced legally enforceable national organic standards.

Before 1991, most other EU member states (except Greece) had a national definition of organic farming overseen by the organic movement and recognized by producers and consumers. In the case of Sweden and the UK, these were also accepted by the government as a basis for policing under food and trading standards (Lampkin *et al.*, 1999a). A growing number of private organic certification marks and labels appeared on organic products in shops across Europe. At the same time, other products were labelled as more environment-friendly, 'green' or 'natural' (even if they did not originate from a different farming system), contributing to increased consumer confusion.

In response to this and in close cooperation with the International Federation of Organic Agriculture Movements (IFOAM) and with the organic movement of some countries (e.g. France), the EU began to draft legislation. EC Reg. 2092/91 defining organic crop production was published in 1991 (European Commission, 1991) and became law in 1993. The main aim was to reduce confusion and fraud and protect both the consumer and producer, and hence assist the development of a common market for organic food. The result was a legally enforceable and officially recognized common standard for organic crop production, certification and labelling in the EU, replacing and underpinning all existing national legal standards relating to organic crop production.

In most areas the EU production standards were similar to the IFOAM Basic Standards (see Chapter 8, this volume, for more details). Through its provisions for imports from non-EU countries, the EC regulation has affected the development of organic standards worldwide. Initially the regulation did not cover livestock, so that national standards remained in effect during a long period of negotiation after which EC Reg. 1804/1999 was passed in 1999 and implemented in 2000. This amendment defined common rules for organic livestock production, while allowing countries to maintain higher national standards for livestock, but not to refuse imports on that basis (European Commission, 1999b; Padel *et al.*, 2004). Since the introduction of EC Reg. 2092/91, more than 20 other amendments have been passed.

Whether legislation defining organic production either nationally or at the European level assisted the development of the organic sector is a matter of debate (Padel *et al.*, 2002a). It is likely that the introduction of a single European or national regulation, as opposed to a variety of standards, reduces consumer confusion and increases confidence among consumers. In Denmark, the fact that the government was involved in this was seen as an important reason for high consumer confidence in organically produced food (Willer, 1998), but in other countries the governments might not be so trusted.

On the other hand, the unifying framework at the EU level left little room for national or regional considerations. EU member states are allowed to have stricter rules only for livestock. Despite the common EU framework, distortion of competition remains a concern, particularly in areas of production that are not yet regulated, such as glasshouse and fish production. The EU included the need for further development and harmonization of the EU regulation in the European Action Plan for organic farming (see Section 6.9), leading to a new European Council Regulation for organic production that was accepted in June 2007 and will come into force in January 2009, when the detailed implementation rules have been finalized (European Commission, 2007). The new regulation aims to further strengthen consumer recognition of organic produce and the common market, as well as to simplify the regulation by setting out objectives and principles and allow for some regional flexibility in implementation under specific, clearly regulated conditions relating to climate or infant industry status.

6.3.2 National and European logos for organic production

To improve consumer confidence and reduce confusion, the development of easily recognizable, common logos may be even more important than legal definitions. Across Europe, private labels originating from organic producer organizations or the retail sector (some financed by a levy on turnover or organic land area) coexist with state-supported labels. Some governments, notably those of France and Denmark, successfully introduced governmental logos together with their national laws defining organic production; these logos are widely recognized and accepted. In contrast, a semi-governmental Austrian bio-logo remains less widely used, and some supermarkets' own labels (e.g. *Ja natürlich* from Billa) are better known by consumers (Michelsen *et al.*, 2001). Other countries, including Germany and the UK, relied on private sector logos. In part, this was not only because of a reluctance to intervene, but also because of a belief that it was desirable to have private sector engagement and competition among certifiers. However, in 2001, after many years of leaving it to the private sector, the German government introduced a now widely used national logo that may be used on all organic food certified according to the EU regulation.

One amendment to EU Reg. 2092/91 (EC Reg. 221/2000) provides for a European logo intended to communicate clearly to consumers the organic character of the product. This logo competes directly with national logos and with well-established private sector logos. Experience overall suggests that the policy involvement in labelling has improved clarity to consumers in most but not all cases.

6.4 Financial Support Programmes for Organic Producers

6.4.1 National programmes before 1994

The positive perceptions of the environmental and other benefits of organic farming among some politicians and policy makers led to the introduction of financial support programmes from the late 1980s, initially in seven European countries. Denmark was the first EU country to introduce financial support for producers during the conversion period, as part of the 1987 law on organic farming. This was followed by a threefold increase in the size of the organic sector (Dubgaard and Holst, 1994; Lampkin and Padel, 1994; Lampkin et al., 1999a). Other Scandinavian countries soon followed. Specific policies to support organic farming were introduced in 1989 on environmental and surplus reduction grounds in Sweden, where agricultural and environmental policies had been more strongly integrated since 1985. (Policies aimed at reducing environmental pollution from manures and fertilizers were introduced in 1985, and in 1986 a levy on all pesticides was introduced with the aim of reducing pesticide use, cutting surplus production and raising funds for research and extension.) This was followed by Finland in 1990. Sweden was the first country to include support for the continuation (maintenance) of organic production beyond the conversion period, recognizing the ongoing environmental benefits of organic production (Liden and Anderson, 1990; Lampkin et al., 1999a).

In 1989 Germany became the first country to make use of the EU's CAP framework to introduce a large-scale support programme for conversion to organic farming. This used the EU's extensification policy (European Commission, 1988), the main aim of which was to reduce agricultural surpluses through extensifying production, either through a quantitative (20%) approach such as setaside or restrictions on livestock numbers, or through a qualitative approach focusing on less intensive production methods that would achieve a similar reduction in output. Conversion to organic production was considered to be consistent with the qualitative production methods approach, although organic farming as such was not mentioned specifically because of the lack of an accepted legal definition at the time. Payments to producers were based on the areas of crops in surplus grown before conversion. A total of 11,248 farms on 377,000 ha received funding, representing a more than sevenfold increase in the land area managed organically between 1988 and 1993. Uptake of the scheme was high on lower-intensity farms, such as in disadvantaged areas and larger farms. A relatively high proportion of holdings did not opt to be certified as organic at the time, as this was not a requirement.

In Austria, organic farming was considered to be an important means to achieve the re-orientation of agricultural policy towards social and ecological goals in preparation for EU accession in 1995. The government provided funds in 1989 and 1990 to help the organic farming organizations with the aim of building up extension and marketing infrastructure before making conversion payments generally available. Pilot projects for conversion payments started in some regions, followed by a national programme in 1991 and a programme for existing organic producers in 1992. Direct payments were seen as necessary to pay producers for the ecological benefits that the market alone would not compensate them for (Posch, 1990; Michelsen et al., 2001).

Outside the EU, Switzerland in particular has a long history of political support for organic farming. The Canton of Bern introduced conversion support in 1988, followed by Baselland in 1989, mainly because of political pressures in their parliaments. In 1992, the Parliament of the Swiss Federation decided to give annual direct payments to farmers, starting in 1993. All cantons paid a lump sum premium (i.e. not an annual payment), one part per farm and another part per hectare, depending on the kind of crops grown; farms had to be certified by the Swiss control organization VSBLO. These rates were set to cover about half of the conversion costs that a farm would incur without access to premiums. Similar to other countries, the uptake was higher in mountainous and hilly areas, but the conversion subsidies did not encourage many arable farmers to convert to organic farming (Schmid, 1994).

6.4.2 EU-wide support for conversion to and continuation of organic production as part of agri-environmental schemes

The 1992 reform of the CAP saw the introduction of an EU-wide agri-environmental support programme (EC Reg. 2078/92), implemented from 1994 (European Commission, 1992). Under this programme, all EU member states had to offer a scheme for grants in aid for converting to or continuing with organic production methods, subject to positive effects on the environment. This organic farming support was one of a range of available agri-environmental measures (e.g. reduced inputs, schemes for specific habitats), and in some cases combinations of different schemes on the same land were also possible. The specific inclusion of organic farming as an agri-environmental scheme was possible because of the EU-wide regulation defining organic farming (EC Reg. 2092/91) adopted the previous year. The support schemes were partially financed by the EU Commission (normally 50%, but 75% in poorer regions), with the balance contributed by the member states or regions (Lampkin et al., 1999b).

Most organic farming schemes under this regulation were introduced in 1994. All countries except France and the UK supported not just the conversion period, but also continuing organic production, usually at a lower rate. In the case of Austria, Posch (1997) claimed that the government did not want to encourage entrants who were interested just in subsidies, so the rate for conversion was kept at the same level as for continuing with organic production. Some countries introduced additional environmental requirements.

In nearly all countries, organic crop production had to be controlled according to EC Reg. 2092/91, but certification of livestock production remained under national supervision because the regulation had not yet been extended to cover livestock (see above). The Swedish case is particularly interesting, because it was the intention of the government to maintain a clear distinction between certified organic production for the market and policy support for organic farming for agri-environmental reasons. Until recently Sweden was the only EU country that had significant areas of non-certified organic land from which no products are sold into the organic market. The producers receiving the organic farming grant are inspected under the same regime as those that receive other agri-environmental support.

Agri-environmental programmes were clearly very important for the development of the organic sector, with the majority of certified holdings and land area in the EU supported under such measures in 1997 (Table 6.1). The most dramatic increases in land area usually took place shortly after the scheme was introduced in each country, followed by periods of consolidation and sometimes even decline. However, uptake of the organic farming options varied widely. This variation can in part be linked to scheme-related factors, such as the level of payment, the existence of a previous organic support scheme and the general popularity of agri-environment schemes.

Table 6.1. Organic farming agreements, area supported and annual expenditure as proportion of all EU agri-environmental support, 1997 and 2003. (From Lampkin et al., 1999a,b; Lampkin and Stolze, 2006.)

	1997[a] (EU15)		2003[b] (EU25)	
	2078/92 organic farming schemes	Share of all 2078/92 agri-environmental schemes (%)	1257/99 organic farming schemes	Share of all 1257/99 agri-environmental schemes (%)
Number of agreements	65,000	4	93,000	5
Land area supported (million ha)	1.3	5	2.9	7
Certified organic area (million ha)	2.1	–	6.2	–
Percentage of certified organic area supported	61	–	47	–
Total annual expenditure (million €)	260	11	500	14
Expenditure (€/ha)	200	2.2[c]	173	2.0[c]

[a]EU15 2078/92.
[b]EU25 1257/1999 and CH (excludes €132 million for 293,000 ha still supported under old 2078/92 agreements).
[c]Ratio of expenditure per hectare in organic farming agreements to all 2078/92 schemes.

In Austria and Finland, for example, the whole agri-environmental programme was a cornerstone of agricultural policy after entry into the EU. Support was claimed on more than 90% of the land area, but only 12.6% and 7.6%, respectively, was spent on organic farming options. In Denmark on the other hand, only 3.9% of all land was in agri-environmental programmes, but most of the money (58.2%) was spent on organic farming options. Factors such as economic pressures, food scares and animal health crises such as foot-and-mouth disease can also influence farmers' propensity to convert, making it difficult to establish clear causal connections to specific aspects of the programmes (Lampkin *et al.*, 1999b; Padel *et al.*, 1999; Dabbert *et al.*, 2004).

The uptake of the schemes also varied within countries, with higher uptake among moderate to low intensity livestock farms, particularly in marginal areas, and among dairy and mixed cropping farms (Schulze Pals *et al.*, 1994; Schneeberger *et al.*, 1997; Michelsen *et al.*, 2001). Specialized crop producers (both arable and horticulture) and intensive pig and poultry producers were less attracted.

The development in some countries has shown that a small proportion of producers who enter organic production when a new scheme is offered will not maintain this approach beyond the time covered by the agreement. For example, in Austria, the organic land area declined at the end of the first 5-year period of the agri-environmental grants in 2000, when approximately 1500 or 8% of the participating farmers went back to conventional production. In surveys, farmers mention a range of reasons for reverting: problems finding a market for their livestock products; bureaucratic aspects of certification; and personal reasons, such as age of the farmer (Kirner and Schneeberger, 1999; Kirner *et al.*, 2005). However, not all countries have experienced a reduction in organic land area at the end of a support period, and some countries that did, or that remained static, are showing signs of growth again in what may be a 4- to 6-year growth and consolidation cycle.

Since 2000, EU agri-environmental schemes have been integrated into the Rural Development Programme (EC Reg. 1257/1999; European Commission, 1999a), which has been progressively implemented in the new EU member states that joined in 2004. All 25 EU member states offered financial support for organic farming under this programme by 2004 (Table 6.1). However, modifications have been made to the schemes to target particular farm types (or to discourage too much uptake by some), and in some cases budgetary constraints and governmental changes have led to temporary discontinuation of support. As the organic sector grows and accounts for increasingly higher proportions of agricultural land use, it is likely that budgetary constraints will lead to pressures to reduce support payments, and possibly to the withdrawal of support in some cases.

6.5 Effects of Mainstream Agricultural Policy on Organic Farming Development

The agri-environmental policies discussed so far are only a minor part of the EU's CAP. By far the greatest expenditure is for the Common Market Organization (CMO) measures, sometimes referred to as Pillar 1 of the CAP (contrasting with the rural development and agri-environmental support, known as Pillar 2). The CMO measures cover the main agricultural commodities, including cereals, milk, red meat, sugar, olives and wine, but their focus has shifted considerably over time. Originally, the intention was to support farm incomes and encourage domestic production (food security was a significant post-war concern) by maintaining agricultural prices through a combination of import levies, intervention buying and export subsidies. However, over time this led to significant problems with overproduction and high (and unpredictable) costs. In addition, the intensification of agricultural production stimulated by these policies was widely blamed for significant negative environmental impacts. The setaside and extensification policies introduced in the late 1980s had little impact on the problem, resulting in the major 'McSharry' reform of the CAP in 1992, which also provided the basis for the EU-wide introduction of the agri-environmental schemes described above.

The 1992 reform represented the first major shift away from the post-war CAP framework. Under Pillar 1, price support mechanisms were progressively converted into direct income payments for particular commodities, with production quotas and setaside introduced to limit the total production of those commodities. Payments were made per hectare for specific crops and per head for specific livestock categories.

The impact that these and subsequent CAP reforms had on organic farming has received relatively little attention, despite the potential for conflict between commodity measures and the agri-environmental measures introduced in the 1992 CAP reform. In many cases, the assumption was made that there is no difference between organic and conventional producers in terms of eligibility, and that any impacts, therefore, were likely to be negligible. On balance, the shift to direct payments under the 1992 reform was largely beneficial for existing organic producers. Subsidy payments for crops were no longer linked to yields, but rather to the areas of crops grown. Organic crop prices did not fall as much as conventional prices as a consequence of the reforms, and since the support payments were calculated on the basis of regional average yields, organic producers with below-average yields benefited. There were also higher payment rates for protein crops (peas and beans) that are used in organic rotations.

However, organic arable producers could not produce the same proportion of supported crops in their rotations because of the need for fertility-building crops; similarly, organic livestock producers had lower stocking rates. Therefore, compared with similar conventional producers, the overall receipts from direct payments were reduced on organic farms by as much as 38% (Häring et al., 2004). The higher levels of agri-environmental support reported in the previous section therefore need to be seen in this context.

Even though organic farmers with their lower yields and livestock densities had not contributed as much to surplus production, all recipients of the direct payments had to agree to set land aside. Given the high levels of market demand for organic products, this requirement meant that some land could not be used for producing organic cash crops that were in demand, while yields per hectare were lower. However, organic producers with little or no livestock could use the setaside payments to generate a return on the fertility-building phase of rotations, especially during conversion.

The adverse impacts of the mainstream measures on organic farms identified above were not necessarily widespread. Organic livestock producers received support for fewer livestock than if they had remained conventional, but benefited from some increases in support payments, such as the higher beef extensification payments for stocking rates less than 1.4 livestock unit (LU) per hectare of forage (1 LU is equivalent to 1 dairy cow). Horticultural crops, grassland and dairy cows were not eligible for support, so organic dairy and horticultural producers, a relatively high proportion of organic producers in most countries, saw few benefits from the reform.

The more negative effect of the 1992 reforms may have been on conventional farmers' willingness to convert to organic farming. Payments differentiated by crop type and livestock eligibility quotas tended to freeze production patterns and levels of intensity. This did not fit well with the enterprise restructuring that typically occurs during conversion to organic farming. In particular, arable farmers were worried about losing eligibility for crop-based payments without getting easy access to some livestock premiums instead. Similarly, converting livestock farmers would receive payments on fewer animals, but were not entitled to arable area payments for any new arable crops introduced, an active disincentive to a more diversified farm structure. On the other hand, the ability to trade support entitlement quotas eased the restructuring process during conversion, and for many producers the ability to lease out quotas during conversion proved to be an important means of financing the conversion.

EU agricultural policy is programmed in 7-year cycles. The 1992 reform, covering 1994–1999, was followed by 'Agenda 2000', covering the period 2000–2006. Although ambitious proposals were made, the political conditions did not result in fundamental changes to the main commodity regimes of the CAP, reinforcing rather than substantially advancing the reforms started in 1992. Measures that were previously advantageous to organic producers remained so. From 2002, the compulsory setaside requirement was lifted from wholly organic holdings, recognizing the lower production levels and high market demand for organic crop products. However, other production constraints, such as quotas, need to be re-examined on similar grounds.

Tensions with the World Trade Organization (WTO) and the budgetary implications of EU enlargement, among other factors, meant that the further fundamental reforms that should have been achieved by 2000 had to be revisited in 2003. At that point, the 'Luxembourg Agreement' resulted in the introduction of 'single farm payments' from 2005, with the expectation that these will remain in place until 2013. This reform involved the conversion of the previous area- or headage-based direct payments into a single payment per holding, i.e. payments are no longer linked to specific commodities. Member states were given a range of choices for implementing the programme, which has meant that the concept of a 'common' European agricultural policy has been diluted. But the impact of these changes on organic producers is still of relevance. Depending on the approach that was chosen, in some countries organic producers (and other participants in agri-environmental measures) are potentially worse off. Recognizing this, the EU introduced some flexibility to moderate the impacts; but where established organic producers have not been able to take advantage of this, they face being left at a competitive disadvantage compared with new converters, who, being more intensive at the time the new policies were introduced, have inherited a higher level of single farm payment per hectare.

6.6 Processing and Marketing Support Programmes

Most conversion aid programmes, both before and under EU-wide agri-environmental support, have led to significant increases in the supply of organic products. It therefore became very important to develop the marketing structure and establish new retail outlets to deal with the supply-led expansion and to maintain premium prices (Hamm and Michelsen, 1996). A number of countries introduced grant schemes for processing and marketing at the national or regional level, through which organic enterprises have received funding.

Denmark and Austria were the first countries to provide support for market development, as early as 1987 and 1989, respectively.

Denmark's law on organic farming from 1987 provided also for market development projects, including collection and processing of organic raw materials and informational and promotional activities. From 1994 onwards, the 'Green Fund' supported environmental and ecological initiatives in urban areas also, and several organic initiatives benefited. From 1996 onward, organic farmers could receive up to 50% of the additional costs, compared with 40% for their conventional counterparts (Lampkin et al., 1999a). The 1989 Agriculture Act in Austria gave special support to organic producer organizations to develop infrastructure and markets. Since 1995, several organic organizations have benefited from special support for investment, wages and consumables available to expand production and meet demand for environment-friendly or low-input foods, including organic foods (Lampkin et al., 1999a,b).

Germany experienced problems of oversupply and an erosion of premiums after the introduction of conversion aid in 1989. However, it was not until 1996 that special guidelines were developed to support the marketing of products labelled according to specific production rules aimed at the organic sector. These were included in the joint Federal Programme for the Improvement of Agricultural Structures and Coastal Protection (*Gemeinschaftsaufgabe Agrarstruktur und Küstenschutz*), under which 52 producer initiatives received support.

At the EU level, several regulations were aimed at strengthening the structure of the markets for agricultural products in general, and organic marketing organizations could apply for this support. Investments relating to organic farming products were identified as priorities for the application of EC Reg. 866/90 on improving the processing and marketing conditions for agricultural products (European Commission, 1990). In eight countries, organic-related activities benefited under this regulation (Lampkin et al., 1999b).

Denmark and to a lesser extent Austria are the only countries that integrated more market-oriented measures with organic production aid schemes right from the start. The example of Denmark illustrates that this can result in the development of a diverse marketing structure, provide help in entering mainstream marketing and help overcome problems such as discontinuity of supply and lack of widespread distribution. In developing action plans (see Section 6.7), several other countries and regions have recognized that organic conversion programmes should be embedded in a more integrated support policy that takes the development of the organic market into account and includes market and information-related measures. The Netherlands, in particular, placed emphasis on the development of supply chain agreements as an alternative to supporting conversion of land directly (Regouin, 2004). However, this had mixed success, because the conversion of pig units was too rapid. Not all the output could be used on the market, which led to an industry-funded compensation schemes for some producers to revert to conventional production (Taen *et al.*, 2004).

Since 2000, there has been increased emphasis at the EU level on support for marketing and processing of organic food to balance the large increases in supply that took place in the 1990s. In particular, support has been provided through the Rural Development Programme (1257/1999), as well as through the structural measures designed to support poorer regions of the EU, and the LEADER programmes to support grass roots initiatives in rural areas (see Section 6.8). Increasing support has also been given to demand-pull measures, such as consumer promotion (also with significant national campaigns in countries like Germany) and prioritization of organic food in public procurement programmes for schools and hospitals (e.g. Rech, 2003).

6.7 Information-related Support Measures

6.7.1 National support for organic information, advice and research

The first country to provide organic information and advice nationally was Switzerland. A national extension programme has been coordinated by FiBL (Research Institute for Organic Agriculture, described in detail in Chapter 14, this volume) since 1977 and has received financial support from the Federal Office of Agriculture and the cantons since 1984.

The case of Austria illustrates that the development of public support for organic extension and advisory services can be very gradual. The normally dominant institutions in the agricultural field, the chambers of agriculture, the cooperatives and the Ministry of Agriculture, initially were quite hostile to the idea of organic farming. Organic producer associations developed their own extension services, receiving some grant aid to do so in the late 1980s. In 1992, regional agricultural chambers began to employ organic advisory staff, in some regions in close cooperation with an organic producers' organization (Ernte). In other regions the chambers advocated organic production as one option among several other agri-environmental schemes, such as reduced inputs. It appears that political intervention from the Minister of Agriculture at the time helped to overcome the conflict between organic and conventional agriculture that had dominated the early years of development, but this may have led to the loss of a distinct identity for organic farming in some regions (Michelsen *et al.*, 2001).

In Denmark, government-supported organic advice began in 1985, and almost from the start was based on close collaboration between organic farmers and one of the two national farmers' unions, which also provided advice. Specialized advisory support was included in the first Danish law on organic farming (1987), and became more integrated with the main advisory system, training a larger number of people to provide organic advice in 1995 while recognizing that specialized skills might be required for conversion planning, although frequently advisors worked for both organic and conventional farmers.

In Germany, also, policy support for organic extension increased region by region throughout the 1980s, with a few organic specialists co-financed by regional governments in most Federal States, depending on the number of organic farms and political alignments. Federal policy initiatives in this area did not start until 2001, when as part of the support programme for organic farming under the Green Minister of Agriculture, an Internet information site was set up (www.oekolandbau.de) with a specialized area for producers. The programme also supported nationwide coordination of advisory activities for specific enterprises and applied research.

Reviews of the development of the organic sector and action plans in some other countries included recommendations to improve the provision of advice to organic farmers, for example in France (Riquois, 1997), Finland (MAF, 1996) and Norway (Landbruksdepartementet, 1995). England and Wales launched an Organic Conversion Information Service (OCIS) that included a telephone helpline and up to 2 days of free advice for those interested in conversion. Outside the EU, Norway introduced a programme of conversion information in 1998 provided by 26 extension rings throughout the country.

Several European countries have supported organic farming research from public funds. In Denmark and Finland, research activities are nationally coordinated by a public research institution; in Switzerland and Norway, the governments channel their spending through private institutes (although the main Norwegian one has recently become part of a larger public agricultural research institution). The UK and Sweden have dedicated public research funds for organic farming. In Spain, organic farming is included in the national agricultural research programme; and the German Bundesprogramm has financed several applied research projects for the benefit of the sector.

6.7.2 European information-related policy support

At the EU level, policy initiatives in the area of information provision were limited to start with. Several collaborative organic farming research projects and networks received funding under the EU Research Framework Programmes since the early 1990s, with a significant increase in the size and number of projects funded since 2000. Recently, there have also been EU-sponsored moves to achieve greater coordination of national research programmes, including direct national funding of transnational research projects, through the CORE-Organic ERAnet established in 2004 ('Coordination of European Transnational Research in Organic Food and Farming'; www.coreorganic.org).

The agri-environment and Rural Development Programmes (EU Regs 2078/92 and 1257/1999) provided for training to support the uptake of the organic and other agri-environmental schemes (European Commission, 1992, 1999a). In addition to the countries that had national extension programmes, the Netherlands, Belgium, Finland, Spain, Portugal, Sweden and some regions in Italy drew up information-related or demonstration measures to improve the uptake of agri-environmental programmes, including organic farming, during the mid-1990s. The Swedish programme explicitly stated an aim to avoid competition with commercial consultancy services, but was open to all organic producers, not only during conversion.

The provision of information and advisory services for organic agriculture received varying degrees of public support but has been patchy throughout Europe. This is in part a reflection of the different structures of general agricultural extension services and the public funding commitment towards them. The development also illustrates differences in the level of collaboration between the organic sector and mainstream agriculture. In the Scandinavian countries, the Netherlands, France, Germany, Austria and Switzerland, the involvement of the general agricultural extension services and public funding for information and providing advice on organic farming has increased considerably. This is likely to have resulted in improved access to information for interested conventional producers, illustrated by the fact that often, considerably more farmers make use of conversion information services than proceed to convert at the time (Midmore *et al.*, 2001). Organic producers, on the other hand, were concerned that the advice given by publicly funded institutions might not be upholding the core principles of organic farming (Gengenbach, 1996; Burton *et al.*, 1997).

A model aiming to overcome such problems was realized in Austria, Germany and Belgium. The main advisory services cooperated closely with organic producer organizations, whereby the supervision of organic advisors would be given to the producer organizations or a joint committee. Similarly, extension rings or organic farmer clubs (e.g. *Ökorings* in Germany) have a membership structure but receive some funding to employ advisors, thus allowing the organic producers to take control of the subjects (Hamm *et al.*, 1996; Luley, 1997). However, membership structures have the drawback that the availability of advice to interested conventional producers may be reduced.

Some countries have given public support to regional and discussion groups of organic producers. Given the importance that converting producers place on seeing good examples (e.g. Wynen, 1990; Burton *et al.*, 1997), the support provided to demonstration farm networks in some countries is of interest. In such a network, experienced organic producers show their farms to visiting groups and receive help with preparing information material, along with some compensation for their time and effort so that they do not have to charge all costs to the visiting groups.

The quality of any information and advice depends on the knowledge and experience of those that provide it. The Austrian government set up a training programme for organic consultants; in Denmark two organic farming specialists in the head office for all advisory services support the field staff in giving organic advice throughout the country. The German programme of 2001 included training sites for new advisors and some support for networking activities. Denmark and Wales have used public funds to support information centres for organic farming, which includes coordination of advisory activities. The EU guidelines for the next Rural Development Programme (European Commission, 2005; see Section 6.8) envisage that advisory services and support to meet community standards and new challenges will be put in place.

6.8 Organic Farming in Rural Development Programmes

Since the mid-1990s it has been recognized that organic farming can help meet many of the goals of regional development programmes, combining a sustainable model of agriculture with the encouragement of local production, processing, and consumption patterns and local marketing networks, thereby increasing the 'economic value' of a region (Vogtmann, 1996). Pilot initiatives for bottom-up regional and rural development were supported by the EU LEADER programme. LEADER I (launched in 1991) supported rural development projects designed and managed by rural associations and local partners. LEADER II (1994 to 1999) sought to encourage models for rural development initiatives at the local level and to support projects that demonstrate new directions for rural development in areas such as environmental protection and quality of life, including transfer of experience and cross-border projects. Organic farming projects with a wide range of aims were supported under LEADER in nine countries during the 1990s (Lampkin *et al.*, 1999b), covering areas such as marketing of regional or specialized organic products, organizations promoting organic agriculture and organic model or demonstration projects, including some linking organic farming and eco-agrotourism. Significant regional development initiatives for organic farming outside EU legislation have occurred since 1997 in the Rhône-Alpes and Pays de Loire regions in France. LEADER initiatives were maintained under Agenda 2000 (2000–2006) and will be fully integrated in the 2007–2013 Rural Development Programme as the fourth axis.

Organic projects have also benefited under the EU structural funds (improving the structure of the agricultural industry) covering a variety of activities, including direct marketing, promotion of regional products, small-scale activities, research, technical advice and training. The schemes have been particularly successful in Germany in developing regional marketing networks, overcoming the problems of a small organic sector, and encouraging the entry of new operators. The impact of grant aid on the organic sector and consequently the development of the region in some cases is significant, as an evaluation of the Irish programme illustrates (Fitzpatrick Associates, 1997).

Since 2000, an integrated approach to rural development has formed the second pillar of the CAP. In the first Rural Development Regulation (EC Reg. 1257/1999), which covered the period 2000–2006, economic development measures were integrated with agri-environmental ones, recognizing the multifunctional nature of agriculture and rural areas, and the contribution that agri-environmental support can make to rural development. The potential benefits of this approach have been realized in a wide range of initiatives (Häring et al., 2005). This process will be continued under the next rural development programming period 2007–2013, with further integration of different measures planned. However, the separate responsibilities of different agencies in implementing specific groups of measures may prove an obstacle to this in practice. Of great significance for the future development of the organic sector is the potential for the Rural Development Regulations to support integrated action plans targeted at the specific, local needs of the organic sector, and to achieve a better balance between supply-push and demand-pull measures.

6.9 Integrated Action Plans and Organic Policy Advisory Bodies

The balancing of supply and demand initiatives to achieve sustainable development of organic agriculture is a key problem facing policy makers seeking to support the organic sector's environmental and rural development goals while retaining its market and consumer orientation. Several countries have developed integrated action plans to achieve a better policy mix (Lampkin, 2003; Lampkin and Stolze, 2006). The range of approaches adopted, however, illustrates the problems and the political pressures inherent in achieving this.

The earliest example is that of Denmark, which not only has a long history of policy support for organic farming, but also adopted an integrated approach from the start. The Danish law of 1987 established the Danish Council of Organic Agriculture, as well as introducing direct payments, national standards and advisory support services. This partnership among the government, organic producer organizations, conventional farming groups, trade unions, and consumer and environmental groups was set up under the Ministry of Agriculture to assess the development possibilities of Danish organic agriculture, develop proposals for further support and assess extension and experimental work. Its work resulted in the first Danish Action Plan (1995–1999), which included a target of 7% of land area to be certified organic by 2000; this was almost achieved (6% of UAA was certified organic by 2000). Action Plan II (MFAF, 1999) aimed for an increase of 150,000 ha, to about 12% of agricultural land, by 2003, although this was not achieved because of oversupply problems, particularly in the dairy sector. The second action plan was characterized by an in-depth analysis of the situation, resulting in 85 detailed recommendations targeting demand and supply, consumption and sales, primary production, quality and health, export opportunities, and institutional and commercial catering.

Other countries followed the Danish lead in developing action plans for the future development of the organic sector. Stolz and Stolze (2006) provide a recent overview of national action plans in different European countries.

In 2001, following an international conference organized by the Danish Ministry (MFAF, 2001), the European Council of Ministers asked the European Commission to develop a European Action Plan (available at: europa.eu.int/comm/agriculture/qual/organic/plan/index_en.htm). This was eventually published in June 2004 and was adopted by the Council of Ministers later that year (European Commission, 2004). The EU Action Plan is based on extensive consultation with member states and stakeholders, and aims to lay down the basis for policy development in future years. The plan recognizes a dual role for organic farming in European policy: supporting delivery of public goods such as environmental protection while also meeting market demands for high quality, safe food. It calls for both aspects to be developed in a balanced way. In all, 21 concrete action points deal with such topics as: consumer information on organic farming (including promotion of the EU logo); streamlining public support via rural development; improving and further harmonizing production standards and inspection regimes; and strengthening research. The action points in relation to streamlining of public support encompass diverse activities: suggestions to national governments to stimulate the demand side through quality assurance schemes; actions to preserve the environmental and nature protection benefits in the long term; developing incentives for organic farmers to convert the whole farm instead of just part; developing incentives for producers to facilitate distribution and marketing; and training and education for all operators involved in organic farming, production, processing and marketing. Of the 21 action points, 15 are related to aspects of the regulations governing the legal definition and control of organic food, which are addressed through the major revision of the EC Reg. 2092/91 initiated in 2005. It is too early to assess the effectiveness of broad framework action plans of this type compared with the more targeted national or regional plans. Lampkin and Stolze (2006) provide a preliminary evaluation of the EU Action Plan; further research is ongoing through the ORGAP project (available at: www.orgap.org), which evaluates the EU organic action plan.

6.10 Discussion and Conclusions

Organic farming has developed rapidly in Europe since the early 1990s, against the background of strong market demand and significant policy support. This support has been both direct, as agri-environmental payments, and indirect, through EU Reg. 2092/91 and related certification activities, as well as marketing and processing activities and information-related measures. This has resulted in a significant shift in perceptions of organic farming. In the 1980s, organic farming was seen as obscure, practised by a few farmers for a minority of consumers who had a preference for such products. By 2000, a strong and growing market for organic products in Europe was leading organic farming from niche to mainstream (Hamm *et al.*, 2002). Michelsen (2001) described this as a breakthrough for organic farming.

Regarding organically managed land area, however, the last few years have seen stagnation or even decline, for example, in Denmark, the UK and Italy. While the circumstances have been different in each case, a significant factor has been the rapid increases in supply, stimulated by both policy and market signals, leading to overproduction and market instability, notably in the dairy sector. However, recent evidence indicates that continued demand growth can turn the situation around, with the dairy sector in the UK and other countries again undersupplied in 2005/06. With revitalized markets and support programmes, countries that had stalled in terms of increasing land area under organic management are now showing signs of growth again, for example, Austria and the UK (Williamson *et al.*, 2006), and more recently, Italy.

The examples show that specific policy support can make a difference in encouraging sector growth, but that interactions with market developments can have significant impacts on policy effectiveness. In this context, the contrast between the development of the organic sector in the USA, which has been primarily market-driven, and the experience in the EU, which has been much more policy driven, is worth closer examination. A first analysis of this has been attempted by US Department of Agriculture researchers (Dimitri and Oberholzer, 2005). The effectiveness of special policies in support of the organic sector can also be hampered through administrative arrangements and through lack of coordination with mainstream agricultural policy. Further research in these areas would be worthwhile.

In examining whether policy support has benefited the organic sector as a whole, two questions appear worthy of consideration: the combined impact on the income of individual organic producers from both direct payments and the impact on prices and markets, and the impact on the integrity of the organic movement and its principles.

The first conversion aid programmes illustrate that the policies were not equally beneficial for all organic producers. There is clear evidence that the grants help compensate for the extra costs in converting to organic production, although these may have varied for different farm types. Also, where the focus was exclusively on conversion, established organic producers were not eligible for the grant or advisory support. They felt disadvantaged and also saw some of their markets and premium prices erode, particularly where the development of consumption did not keep up with rapid increases in production, leading to problems of oversupply in existing markets. This was the case with the first programme of organic farming support in Germany in 1989, leading to an oversupply of organic cereals and erosion of prices. Similarly, average farmgate prices for organic milk were almost halved in England in 2001, 24 months after the reopening of the conversion aid scheme under EC Reg. 2078/92. Significant improvements in conversion support combined with very high organic milk prices attracted many producers into the organic sector. The supply of organic milk almost doubled, and the market could not absorb such sudden increases (Padel et al., 2002b). The situation was made worse by poor market intelligence and lack of forecasting of expected production volumes as well as administrative arrangements, as the scheme had been closed for some time in 1998/99 while payment rates were reviewed.

This illustrates that short application periods and lack of continuity of policy support can exacerbate the problems caused by changes in market and policy support signals. If all producers who might have been considering conversion postpone their decision until an awaited programme opens, many of their products will enter the organic market at the same time, following the statutory conversion period. On the other hand, a country like Denmark illustrates that through wide consultation with the industry and integration of various policy measures, the negative effects can be reduced, although not completely avoided, as strong market signals can also result in rapid supply increases. The Dutch experience with supply chain agreements, designed specifically to avoid the effect of distorting markets through supply-push incentives, also showed that an approach based predominantly on market signals can still lead to problems of oversupply.

In countries where samples of organic farms are regularly monitored as part of farm income reporting, there is evidence that organic producers' farm incomes have benefited from organic aid programmes (Offermann and Nieberg, 2000; Häring et al., 2004; Nieberg et al., 2005; Jackson and Lampkin, 2006). However, there are clear differences among farm types. In the EU, the uptake of the organic options under the agri-environmental programme has been considerably higher among livestock producers than crop producers, particularly in the less-favoured agricultural regions that mainly have dairy, beef and sheep production, where the costs of conversion and extent of necessary technical changes are perceived to be lower. This resulted in an unbalanced market for ruminant livestock products (beef, sheep, goats and milk), in which the supply was higher than demand and farmgate premiums on average lower than for crops (Hamm and Gronefeld, 2004), reducing the overall financial benefit of organic management. In addition, undersupply of arable crops for livestock feed and direct competition with human demand can result in high feed costs. Other producers, for example, in horticulture, have benefited much less from organic programmes in most countries because of a lack of targeted support and their traditional exclusion from most EU Pillar 1 measures.

Overall, it appears that many, but not all, organic producers have benefited financially from policy support. Other benefits to the individual producer are more difficult to quantify and have not been researched intensively, but arguably, producers have benefited from the better access to markets and better availability of information in countries where support in these areas was available.

Within organic organizations the opinion is widely held that the farms converting with the help of policy support have done so for the 'wrong' reasons. It is frequently argued that the pioneers had idealistic motives, whereas the new entrants convert largely because of better expected economic returns, and it is widely assumed that later converters are not so committed to the principles of organic production.

Surveys comparing the motives of earlier and later organic producers in a rigorous way are rare, but there is some indication that the motivation for conversion and the attitudes of organic producers have changed over time (Padel, 2001). Midmore et al. (2001) found that the availability of financial support was one among several reasons to consider organic conversion. Michelsen and Rasmussen (2003) found higher prevalence of utilitarian motives among later converters, but highlighted that, like earlier converters, they also strongly identified with organic values despite the observed differences. In contrast, Lund *et al.* (2002) saw later entrants as having a more superficial relationship to organic principles. However, apart from attitudes towards organic farming and its principles, other factors are also likely to influence organic producers' motives and values, such as professional background, farm type and external economic circumstances (Padel, 2006). Particularly the economic environment for agriculture in Europe and the developed world has changed, making it vital for commercial producers to consider the financial impact of any major decisions, such as conversion. On balance, there appears to be no conclusive evidence that most producers who have converted with the help of grant money are less committed to organic farming and its principles. Some have left support programmes and certification, as in Austria, for example, but whether they continue to farm at all, and if so, in what way, is not clear.

Policy programmes supporting conversion to organic management should attract new entrants into the industry. If they do not, they have failed in their primary aim. The uptake of the various schemes illustrates considerable variation in the success in this respect among countries. However, whether high uptake is good or bad for the organic organizations depends to some extent on the sector's point of view. Clearly, more producers means more competition if existing markets cannot be expanded. However, we should question critically whether a market for certified organic products is an aim of the organic movement in itself, or rather a means to an end, that of achieving health and sustainability in food production systems. Furthermore, the level of grant aid is important in this context. If set at a level that relies on producers achieving a return from the market, the absence of a sufficient market would be a reason to not continue with organic management. Problems with finding a market were seen as an important driving force behind producers considering reverting to conventional production in surveys in Austria and the UK (ADAS, 2004; Kirner *et al.*, 2005). However, the overall proportion of farmers leaving organic farming schemes is relatively low in most countries, taking account of normal retirement rates from agriculture. The Swedish experience shows that support programmes can also be set up purely as agri-environmental measures without requiring certification or assuming that products are marketed as organic.

Observers have also questioned whether the organic movement has lost control over its own destiny (Dabbert *et al.*, 2004), a theme that runs through this book (e.g. Chapters 3, 8 and 16, this volume). Tovey (1997), for example, has argued that EU policy support pushed organic farming towards environmental conservationism rather than sustainable food production. The importance that agri-environmental measures and programmes have had in supporting the organic sector would favour this argument. In addition, in most countries policy intervention has strengthened collaboration between organic farming and general agricultural institutions. Although this helped overcome the very contentious mood that had dominated the early years of development, therein lies the danger of losing a distinct organic identity, and in some regions organic farming has indeed just become one of many agri-environmental schemes. However, in recent years there has been a clear shift in European agricultural policy towards quality assurance considering the whole food chain from 'farm to fork', and organic farming is frequently quoted as one example of this.

The definition of organic farming and the standards in Europe are now part of legislation and therefore no longer fully under the control of the movement. Arguably, by regulating and supporting organic farming, the sector was helped in clarifying its positions (Scharpé, 2001). Government intervention has also led to some tightening of the organic standards, which could go so far as to make it very difficult to farm organically. On the other hand, because the process involves representation from all member states, the EU organic regulation can no longer be changed rapidly in direct response to pressure from lobby groups. As part of the European Action Plan, the EU Commission has recognized the need to further clarify the principles of organic farming, a process now under way as part of EU-funded research (available at: www.organic-revision.org) in parallel with initiatives of IFOAM.

The organic sector in Europe has clearly been influenced considerably by policy support, and over time the organic movement has realized the potential but also some of the dangers. One consequence is that organic organizations in Europe now take political lobbying very seriously. In some countries, policy makers have been aware of the pivotal role of organic organizations in maintaining integrity and credibility, and have made consultation with organic stakeholders, such as producers' organizations and certification bodies, a permanent feature in policy development, as with the Danish Organic Farming Council. The organic sector appears to have been successful in making its case. The agreed principles of the CAP reform of 2004 appear to provide the basis for further growth of the European organic sector, possibly aiding the move from a niche to a part of mainstream European agriculture. Such a move would have been very unlikely without governmental support.

Acknowledgement

We gratefully acknowledge financial support from the Commission of the European Communities for two projects on which this chapter draws: Effects of the CAP Reform and possible further development on organic farming in the EU (Fair3-CT96-1794) and EU-CEE-OFP: Further Development of Organic Farming Policy in Europe with particular emphasis on EU Enlargement (QLK5-2002-00917). The chapter does not necessarily reflect the Commission's views and in no way anticipates the Commission's future policy in this area.

References

ADAS (2004) *Farmers' Voice 2004 Summary Report: Organic Farming.* ADAS Market and Policy Research, Wolverhampton, UK.

Burton, M., Rigby, D. and Young, T. (1997) Why do UK organic horticultural producers adopt organic techniques? *NENOF* 1997(6), 7–10.

Dabbert, S., Zanoli, R. and Lampkin, N.H. (2001) Elements of a European Action Plan for Organic Farming. In: *European Conference 'Organic Food and Farming – Towards Partnership and Action in Europe', May 2001.* Ministry of Food, Agriculture and Fisheries, Copenhagen, Denmark.

Dabbert, S., Häring, A.M. and Zanoli, R. (2004) *Organic Farming: Policies and Prospects.* Zed Books, London.

Dimitri, C. and Oberholzer, L. (2005) *Market-Led Versus Government-Facilitated Growth: Development of the U.S. and EU Organic Agricultural Sectors.* United States Department of Agriculture, Washington, DC.

Dubgaard, A. and Holst, H. (1994) Policy issues and impacts of government assistance for conversion to organic farming: the Danish experience. In: Lampkin, N.H. and Padel, S. (eds) *The Economics of Organic Farming.* CAB International, Wallingford, UK, pp. 383–392.

European Commission (1988) Commission Regulation (EEC) No 4115/88 laying down detailed rules for applying the aid scheme to promote the extensification of production. *Official Journal of the European Communities* L361 (December), 13–17.

European Commission (1990) Council Regulation (EEC) No 866/90 on improving the processing and marketing conditions of agricultural products. *Official Journal of the European Communities* L91(April), 1–6.

European Commission (1991) Council Regulation (EEC) No 2092/91 of 24 June 1991 on organic production of agricultural products and indications referring thereto on agricultural products and foodstuffs. *Official Journal of the European Communities* L198(July), 1–15.

European Commission (1992) Council Regulation (EEC) No 2078/92 of 30 June 1992 on agricultural production methods compatible with the requirements of the protection of the environment and the maintenance of the countryside. *Official Journal of the European Communities* L215(July), 85–90.

European Commission (1999a) Council Regulation (EC) No 1257/1999 of 17 May 1999 on support for rural development from the European Agricultural Guidance and Guarantee Fund (EAGGF) and amending and repealing certain Regulations. *Official Journal of the European Communities* L 160(June), 80–100.

European Commission (1999b) Council Regulation (EC) No 1804/1999 of 19 July 1999 supplementing Regulation (EEC) No. 2092/91 on organic crop production of agricultural products and indications referring thereto on agricultural products and foodstuffs to include livestock production. *Official Journal of the European Communities* L222(August), 1–28.

European Commission (2004) European Action Plan for Organic Food and Farming. Brussels 10.06.2004. COM(2004)415 final. Commission of the European Communities, Brussels. Available at: europa.eu.int/comm/agriculture/qual/organic/plan/comm_en.pdf

European Commission (2005) Council Regulation (EC) No 1698/2005 of 20 September 2005 on support for rural development by the European Agricultural Fund for Rural Development (EAFRD). *Official Journal of the European Union* L227(October), 1–40.

European Commission (2007) Council Regulation on organic prodution and labelling of organic products and repealing Regulation (EEC) No 2092/91. Commission of the European Communities, Brussels.

Fitzpatrick Associates (1997) Mid-term evaluation: development of organic farming measure 1.3 (e). Fitzpatrick Associates, Economic Consultants, Dublin.

Gengenbach, H. (1996) Fachberatung biologisch-dynamischer Landbau in Hessen. *Lebendige Erde* 1996(3), 237–243.

Hamm, U. and Gronefeld, F. (2004) *The European Market for Organic Food: Revised and Updated Analysis*. Organic Marketing Initiatives and Rural Development Series, Vol. 5. School of Business and Management, University of Wales, Aberystwyth, UK.

Hamm, U. and Michelsen, J. (1996) Organic agriculture in a market economy – perspectives from Germany and Denmark. In: Østergaard, T.V. (ed.) *Fundamentals of Organic Agriculture*, Vol. 1, Proceedings of the 11th IFOAM International Conference, August 11–15, 1996, Copenhagen. International Federation of Organic Agriculture Movements, Tholey-Theley, Germany, pp. 208–222.

Hamm, U., Poehls, A. and Schmidt, J. (1996) *Analyse der Beratung von ökologisch wirtschaftenden Landwirten in Mecklenburg-Vorpommern*. Fachhochschule Neubrandenburg, Fachbereich Agrarwirtschaft und Landespflege, Neubrandenburg, Germany.

Hamm, U., Gronefeld, F. and Halpin, D. (2002) *Analysis of the European Market for Organic Food*. Organic Marketing Initiatives and Rural Development Series, Vol. 1. School of Business and Management, University of Wales, Aberystwyth, UK.

Häring, A.M., Dabbert, S., Auerbacher, J., Bichler, B., Eichert, C., Gambelli, D., Lampkin, N., Offermann, F., Olmos, S., Tuson, J. and Zanoli, R. (2004) *Organic Farming and Measures of European Policy*. University of Hohenheim, Hohenheim, Germany.

Häring, A.M., Stolze, M., Zanoli, R., Vairo, D. and Dabbert, S. (2005) The potential of the new EU Rural Development Programme in supporting organic farming. EU-CEE-OFP Project Discussion Paper. Fachhochschule Eberswalde, Eberswalde, Germany.

Jackson, A. and Lampkin, N. (2006) *Organic Farm Incomes in England and Wales 2004/05*. Institute of Rural Sciences, University of Wales, Aberystwyth, UK.

Kirner, L. and Schneeberger, W. (1999) Hemmfaktoren einer Ausweitung des

Biologischen Landbaus in Oesterreich. *Die Bodenkultur* 50, 227–234.

Kirner, L., Vogel, S. and Schneeberger, W. (2005) Ausstiegsabsichten und tatsächliche Ausstiegsgründe von Biobauern und Biobäuerinnen in Österreich – Analyse von Befragungsergebnissen. In: Hess, J. and Rahmann, G. (eds) *Ende der Nische: Beiträge zur 8. Wissenschaftstagung Ökologischer Landbau.* Kassel University Press, Kassel, Germany, pp. 429–432.

Lampkin, N. (2003) From conversion payments to integrated action plans in the European Union. In: *Organic Agriculture: Sustainability, Markets and Policies.* OECD/CAB International, Paris/Wallingford, UK, pp. 313–328.

Lampkin, N.H. and Padel, S. (1994) Organic farming and agricultural policy in Western Europe: an overview. In: Lampkin, N.H. and Padel, S. (eds) *The Economics of Organic Farming.* CAB International, Wallingford, UK, pp. 437–456.

Lampkin, N. and Stolze, M. (2006) European Action Plan for Organic Food and Farming. *Law, Science and Policy* 3, 59–73.

Lampkin, N., Foster, C. and Padel, S. (1999a) *The Policy and Regulatory Environment for Organic Farming In Europe: Country Reports.* University of Hohenheim, Hohenheim, Germany.

Lampkin, N.H., Foster, C., Padel, S. and Midmore, P. (1999b) *The Policy and Regulatory Environment for Organic Farming in Europe.* Organic Farming in Europe: Economics and Policy, Vol. 1, University of Hohenheim, Hohenheim, Germany.

Landbruksdepartementet (1995) Handlingsplan for videre utvikling av økologisk landbruk. Ministry of Agriculture, Oslo.

Liden, C.J. and Anderson, R. (1990) Legislation and measures for the solving of environmental problems resulting from agricultural practices, their economic consequences and the impact on agrarian structures and farm rationalisation. In: Besson, J.-M. (ed.) *Biological Farming in Europe.* Food and Agriculture Organization, Rome, pp. 259–272.

Luley, H. (1997) Die Informations- und Beratungsleistungen von Anbauverbänden und Öko-Beratungsringen zwischen 1988 und 1995. In: Köpke, U. and Eisele, J.A. (eds) *4. Wissenschaftstagung zum Ökologischen Landbau, 3–4 March 1997.* Verlag Dr. Köster, Bonn, Germany, pp. 530–536.

Lund, V., Hemlin, S. and Lockeretz, W. (2002) Organic livestock production as viewed by Swedish farmers and organic initiators. *Agriculture and Human Values* 19, 255–268.

MAF (1996) *Development of organic farming.* Ministry of Agriculture and Forestry, Helsinki, Finland.

MFAF (1999) *Action Plan II-Development in Organic Agriculture.* Danish Ministry of Food, Agriculture and Fisheries, Copenhagen.

MFAF (2001) *Organic Food and Farming: Towards Partnership and Action in Europe.* Danish Ministry of Food, Agriculture and Fisheries, Copenhagen, Denmark. Available at: www.fvm.dk/konferencer/organic_food_farming

Michelsen, J. (2001) Recent development and political acceptances of organic farming in Europe. *Sociologia Ruralis* 41, 3–19.

Michelsen, J. and Rasmussen, H. (2003) *Nyomlagte danske oekologiske jordbrugere 1998. En beskrivelse basered paa en spoersgekemaundersoegelse.* University of Southern Denmark, Esbjerg, Denmark.

Michelsen, J., Lynggard, K., Padel, S. and Foster, C. (2001) *Organic Farming Development and Agricultural Institutions in Europe: A Study of Six*

Countries. University of Hohenheim, Hohenheim, Germany.

Midmore, P., Padel, S., Mccalman, H., Isherwood, J., Fowler, S. and Lampkin, N. (2001) *Attitudes Towards Conversion to Organic Production Systems: a Study of Farmers in England*. Institute of Rural Studies, University of Wales, Aberystwyth, UK.

MLNV (2000) *Evaluatie Plan Van Aanpak Biologische*. Ministry of Agriculture, Nature Management and Fisheries, The Hague.

Nieberg, H., Offermann, F., Zander, K. and Jägersberg, P. (2005) Further Development of Organic Farming Policy in Europe, with Particular Emphasis on EU Enlargement D12: Report on the Farm Level Economic Impacts of OFP and Agenda 2000 Implementation. EU CEE OFP European Organic Farming Policy QLK5-2002-00917, Bundesforschungsanstalt für Landwirtschaft (FAL), Braunschweig, Germany.

OECD (2003) *Organic Agriculture: Sustainability, Markets and Policies*. OECD/CAB International, Paris/Wallingford, UK.

Offermann, F. and Nieberg, H. (2000) *Economic Performance of Organic Farms in Europe*. University of Hohenheim, Hohenheim, Germany.

Olmos, S. and Lampkin, N.H. (2005) Statistical report on the development of OF in EU member states and Switzerland for the period 1997–2002 with update for 2003. Institute of Rural Sciences, University of Wales, Aberystwyth, UK.

Padel, S. (2001) Conversion to organic farming: a typical example of the diffusion of an innovation. *Sociologia Ruralis* 41, 49–61.

Padel, S. (2006) Values of established and converting organic producers: results of a European focus group study. *International Journal of Agricultural Resources, Governance and Ecology* (in press).

Padel, S., Lampkin, N. and Foster, C. (1999) Influence of policy support on the development of organic farming in the European Union. *International Planning Studies* 4, 303–315.

Padel, S., Lampkin, N.H., Dabbert, S. and Foster, C. (2002a) Organic farming policy in the European Union. In: Hall, D. (ed.) *Advances in the Economics of Environmental Resources*, Vol. 4. Elsevier Science, Oxford, pp. 169–194.

Padel, S., Shirley, C., Mayfield, A., Edwards, J., Harward, R. and Stocker, P. (2002b) *Market Prospects for Organic Milk*. MDC Milk Development Councils, Cirencester, UK.

Padel, S., Schmid, O. and Lund, V. (2004) Organic livestock standards. In: Vaarst, M., Roderick, S., Lund, V. and Lockeretz, W. (eds) *Animal Health and Welfare in Organic Agriculture*. CAB International, Wallingford, UK, pp. 57–72.

Posch, A. (1990) *Support for Organic Farming in Austria*. Federal Ministry of Agriculture and Forestry, Vienna.

Posch, A. (1997) Making growth in organic trade a priority. In: Maxted-Frost, T. (ed.) *The Future Agenda for Organic Trade*. Proceedings of the 5th IFOAM Conference for Trade in Organic Products. IFOAM, Tholey-Theley, Germany, pp. 9–12.

Praznan, J. and Koutna, K. (2004) *Development of Organic Farming and the Policy Environment in Central and Eastern European Accession States 1997–2002*. Institute of Rural Sciences, University of Wales, Aberystwyth, UK.

Rech, T. (2003) Organic food for public institutions. In: *Organic Agriculture: Sustainability, Markets and Policies*.

OECD/CAB International, Paris/Wallingford, UK, pp. 401–406.
Regouin, E. (2004) To convert or not convert to organic farming. In: *Organic Agriculture: Sustainability, Markets and Policies*. OECD/CAB International, Paris/Wallingford, UK, pp. 227–235.
Riquois, A. (1997) Pour une agriculture biologique au coeur de l'agriculture francaise: proposition pour un plan pluriannuel de developpement 1998/2002. Ministère de l'Agriculture et de la Péche and Conseil General du Genie Rural des Eaux et Forêts. Paris.
Scharpé, A. (2001) Conference Speech. In: *Organic Food and Farming: Towards Partnership and Action in Europe*. Ministry of Food, Agriculture and Fisheries, Copenhagen, pp. 100–104.
Schmid, O. (1994) Agricultural policy and impacts of national and regional government assistance for conversion to organic farming in Switzerland. In: Lampkin, N.H. and Padel, S. (eds) *The Economics of Organic Farming*. CAB International, Wallingford, UK, pp. 393–408.
Schneeberger, W., Eder, M. and Posch, A. (1997) Strukturanalyse der Biobetriebe in Österreich. *Der Förderungsdienst – Spezial*, 45, 1–12.
Schulze Pals, L., Braun, J. and Dabbert, S. (1994) Financial assistance to organic farming in Germany as part of the EC extensification programme. In: Lampkin, N.H. and Padel, S. (eds) *Economics of Organic Farming*. CAB International, Wallingford, UK, pp. 411–434.
Stolz, M.H. and Stolze, M. (2006) Comparison of action plans for organic agriculture in the European Union. In: Andreasen, C.B., Elsgaard, L., Søndergaard Sørensen, L. and Hansen, G. (eds) *Proceedings of the European Joint Organic Congress, 30–31 May 2006, Odense, Denmark*, pp. 90–91.
Taen, R.J.M., Dalen, L.H.V. and Ruijter, P.P.M.D. (2004) Biologisch meer gangbaar. Evaluatie-onderzoek. Nota Biologische Landbouw 2001–2004. Onderdeel extern onderzoek.
Tovey, H. (1997) Food, environmentalism and rural sociology: on the organic farming movement in Ireland. *Sociologia Ruralis* 37, 22–37.
Vogtmann, H. (1996) Regionale Wirtschaftskreisläufe – Perspektiven und Programme für die Landwirtschaft in Hessen. Für den ländlichen Raum und seine Menschen: 9. Tagung der Landessynode der EKKW. Evangelische Akademie Hofgeismar, Hofgeismar, Germany.
Willer, H. (ed.) (1998) *Ökologischer Landbau in Europa – Perspektiven und Berichte aus den Ländern der EU und den EFTA Staaten*, Deukalion Verlag, Holm, Germany.
Willer, H. and Yussefi, M. (eds) (2006) *The World of Organic Agriculture – Statistics and Emerging Trends 2006*. International Federation of Organic Agriculture Movements, Bonn, Germany.
Williamson, S., Cleeton, J. and Nettleship, T. (2006) *Organic Market Report 2006*. Soil Association, Bristol, UK.
Wynen, E. (1990) Sustainable and conventional agriculture in southeastern Australia – a comparison. Economics Research Report No. 90.1. School of Economics and Commerce, La Trobe University, Bundoora. Available at: www.elspl.com.au/abstracts/EOA.HTM

7 The Organic Market

J. Aschemann,[1] U. Hamm,[1] S. Naspetti[2] and R. Zanoli[2]

[1]*Department of Agricultural and Food Marketing, Faculty of Organic Agricultural Sciences, University of Kassel, Steinstrasse 19, 37213 Witzenhausen, Germany;* [2]*Department DIIGA, Faculty of Engineering, Polytechnic University of Marche, Via Brecce Bianche, 60131 Ancona, Italy*

7.1 The Beginnings of the Organic Food Market

The cradle of organic farming lies in Europe, where different actors experimented with and developed forms of alternative agriculture from the early 1920s (as discussed in more detail in Chapter 2, this volume). A worldwide organic market has existed ever since. For example, the first organic coffee farm was founded in 1928 in Mexico, and the first organic logo ever was the biodynamic 'Demeter' logo introduced the same year (Demeter, 2006).

Only after World War II did the differences between organic food and that produced by 'conventional' industrialized farming systems become more evident to consumers. By the 1960s, a range of alternative food distribution networks – including non-profit food cooperatives and communes – had also been established, providing for the local distribution of organic food (Hamm and Michelsen, 1996). In the early 1970s, environmental movements increased their focus on organic farming, and consumers were specifically targeted. In 1972, IFOAM was founded (for details see Chapter 9, this volume), and in 1974 Rapunzel Naturkost (now one of the largest organic wholesalers in Germany) was established. In those years one goal of the organic movement was to encourage consumption of locally grown food, and indeed the first organic consumers in many countries of Europe organized themselves in groups to provide their families with a regular supply of local, 'safe' foods. However, not all of these were truly organic, given the lack of official organic regulations in many countries.

7.2 Worldwide Production, Consumption and Trade

In the 1970s, organic farming started to become a widespread phenomenon when consumer awareness of environmental and health issues grew in Europe, North America and Japan, leading to a willingness to pay premium prices for organic foods. In the 1980s and 1990s, when standards were set and some governments introduced organic aid schemes for farmers, organic farming became officially recognized. A common definition of organic production and the recognition and traceability of its products through a legal framework of logos and certification systems were the prerequisites for the rapidly growing international market in the 1980s and 1990s. Growing consumer interest, partly induced by food scandals, led to the engagement of major food retailers, and this in turn to a growth of the market through greater availability and recognition of the products. As more and more consumers in Europe and the USA wanted to buy a full range of food products of organic origin, major distributors looked for suppliers of tropical and out-of-season products. On account of this, developing countries increasingly contributed to organic production and today show the highest growth rate in area farmed organically, especially in Latin America (Willer and Yussefi, 2006). Still, most products of developing countries are destined for export to the major markets, which are the industrialized countries of Europe, North America and Asia. The possibility for countries in the southern hemisphere to grow crops out-of-season for the large markets in the northern hemisphere has also led countries such as Argentina, Chile, South Africa, Australia and New Zealand to be highly involved in the worldwide organic trade.

A slowdown of the growth in the organic market has been observed since the turn of the millennium. This has been interpreted either as a sign that the market has reached maturity, or, as argued by most market observers, as the result of the overall economic recession. The slowdown has especially occurred in Europe, but also in Japan because of the introduction of a government regulation that temporarily caused many organic products to lose their status. The organic market in North America developed with a time lag, but around 2000 overtook the European market, the previous leader in volume (Willer and Yussefi, 2006), for a short period of time. This was reversed recently with the declining value of the US dollar relative to European currencies. After 2003, with the Euro becoming stronger, countries in the southern hemisphere and developing countries were more interested in exporting their organic products to Europe. Nevertheless, the US market is expected to have high growth rates in the future because of strong consumer demand.

Since 2000, the Stiftung Ökologie und Landbau has been gathering data about organic farming on a worldwide level, with vital market data

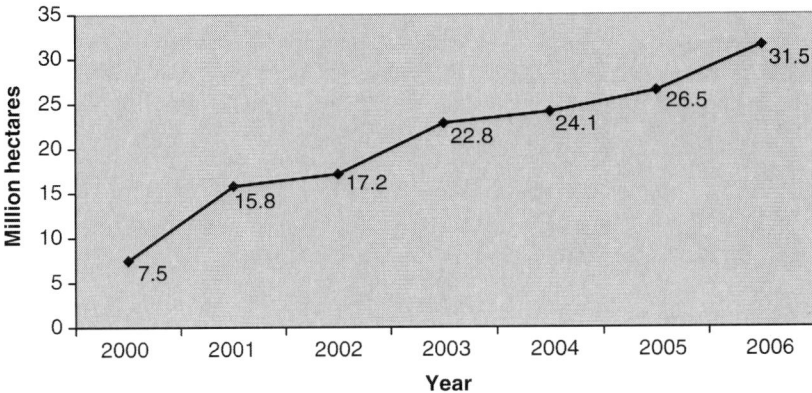

Fig. 7.1. Development of land area under organic management, 2000–2006. (From Willer and Yussefi, 2000, 2001, 2004, 2005, 2006; Yussefi and Willer, 2002, 2003.)

originating in the International Trade Centre and IFOAM members. When interpreting the data, it must be kept in mind that many data are based on estimates and that part of the growth of organically managed area may also be the result of increased data availability, especially in developing countries. As can be seen in Figs 7.1 and 7.2, both the organic area and the organic market show an impressive growth.

Tables 7.1 and 7.2 show the share of the total organic area and the market revenues for the different continents. Obviously, the consumer market lies in the industrialized countries, whereas a major part of the production area lies in Oceania and Latin America. It must be taken

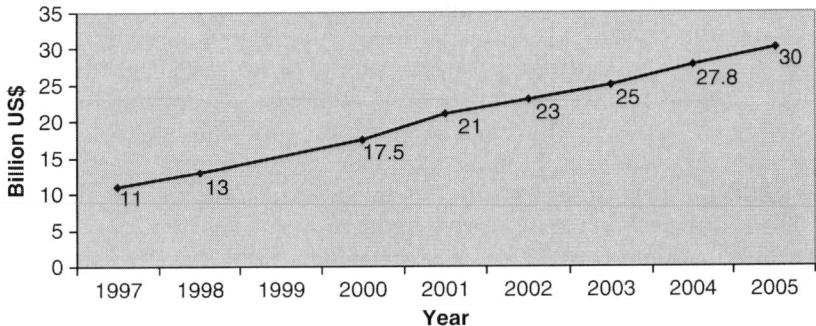

Fig. 7.2. World market volume of organic agricultural products, 1997–2005. (From Sahota, 2006; Willer and Yussefi, 2000, 2004, 2005; Yussefi and Willer, 2003.)

Table 7.1. Area under organic management: % of total for each continent in 2005. (From Willer and Yussefi, 2006.)

Oceania	39
Latin America	20
Europe	21
North America	4
Asia	13
Africa	3

Table 7.2. Distribution of global organic food and drink revenues (%), 2004. (From Sahota, 2006.)

North America	47
Europe	49
Others	4

into account that the latter areas are under particularly extensive management. North America currently has especially high import rates in comparison to Europe, which is due to governmental support of the organic supply side in Europe and a recent boost in demand in the USA and Canada.

7.3 Development of Marketing Channels

In the beginning of the organic movement in the 1920s, marketing of organic products took place in small and often close private networks of farmers and consumers. Organic farmers who were convinced that they had found the right production methods looked for consumers willing to pay higher prices for their special products. Direct marketing of organic products to consumers was the main marketing channel for most organic farmers before World War II. In some countries, especially Germany and Switzerland, health food stores (*Reformhäuser*) and a few shops specializing in fruits and vegetables started to sell organic plant products in the 1930s. However, distribution systems and logistics were not well developed, as stated by a German producers' organization, Arbeitsgemeinschaft Landreform, in 1938:

> It is unfortunately a fact that the handling…especially of vegetables and fruit from the farmer to the consumer is so bad that the condition of organic products does not satisfy the normal requirements at all. Some shops,

vegetable shops as well as food and health stores, are not adequately equipped for keeping vegetables fresh and their managers are not trained sufficiently for this purpose. A first prerequisite for the sale of organic products is to supply the consumer with fresh goods. This, however, in itself has several very important requirements. The quantity must be large enough, it must be available in all seasons, the place of production must make quick transport available to a central point from which distribution can take place quickly and without a hitch. One farm can seldom supply sufficient goods and, above all, supply them evenly over the whole year. Therefore, many farms in one area, on the periphery of a town, must focus on organic production and be integrated into standardized production and supply. Every organic farmer should be interested in having as many neighbouring farmers as possible convert to organic production. Labour and costs for transport will be reduced to a minimum by this. Selling products in the town is then the least of one's worries if the goods are delivered fresh and regularly in sufficient quantities.
(Translated from Arbeitsgemeinschaft Landreform, 1938, p. 32)

The problems stated in this 1938 article are characteristic of the start-up phase of nearly all organic markets worldwide.

After World War II, direct marketing of organic products from farmers to consumers was still the most important channel. In some countries, such as Germany, Switzerland and the UK, producer–consumer associations were founded and the distribution system was improved. In the late 1950s and 1960s, additional farmers' marketing associations were built up. They looked for processors and found mostly smaller processors willing to become involved in processing and marketing organic food. These products were mainly sold through health food stores and smaller general food retailers. Because of the small amounts, distribution costs were very high, raising retail prices far above the prices of ordinary foods. Therefore, organic food was able to gain only a marginal market position, with market shares below 0.1% until the early 1980s.

In the 1970s the first specialized retail shops for organic products were founded in many central European countries and later in the USA and Japan. The people running these shops and their clients often originated in the 'green' or alternative movements (Böckenhoff and Hamm, 1983). People in these movements cared not only about environmental problems arising from the industrialization of agriculture, but also about problems involving nuclear power stations, industrial pollution of air, water and the soil, and so forth. Still other interests of the alternative movement came out of a global environmental perspective (e.g. 'save the rainforest') as well as concern over social justice in developing countries (e.g. fair trade, support for small farmers or for the populations of countries with repressive governments, such as El Salvador and Nicaragua). As varied as have been the concerns of the clients and owners of these specialized retail shops, so too has been the range of products offered: organic products, fair trade products, products from alternative cooperatives in special regions and so forth. As the number of these alternative or green shops grew rapidly, the first wholesalers developed from retail shops.

Retail shops started to get specialized in most central European countries, and in some countries, such as Germany and the Netherlands, shops selling only organic foods became the most important sales channel for organic products in the 1980s. In Mediterranean countries, North America, Japan and New Zealand, total specialization of alternative food shops in organic foods was uncommon. Most sold organic foods, together with 'wild-grown' herbs or fruits, herbal teas and other items that were not necessarily of organic origin but met the demands of their alternative clients.

As the supply grew significantly in Germany, France and Italy in the early to mid-1980s, farmers and processors were confronted with sales problems (Hamm, 1986). The small organic food shops and direct marketing by farmers were not able to increase their organic sales as quickly as supply grew. Consequently, a new sales channel for organic products was opened: conventional supermarkets. Amid heated debates on the opportunities and threats from big conventional supermarket chains getting into the organic market (Hamm, 1986), some organic farmers' cooperatives in Germany started to deliver to supermarket chains in the 1980s. The small but quickly growing organic market and the high premium prices for organic products induced several conventional supermarkets, smaller as well as larger, to start offering organic products both in Germany (e.g. tegut, Kaufmarkt, Coop, VeGe, Nanz, Tengelmann) and in the UK (e.g. Marks and Spencer, Sainsbury's). In the late 1980s many conventional supermarket chains throughout central and northern Europe followed the pioneers, but with very different success. Supply was not sufficient to simultaneously serve several big supermarket chains with organic products of satisfactory quality, and these chains' buyers had many problems with the few organic products and many different suppliers. At the same time, the sales personnel of the supermarket chains also had problems dealing with the flood of questions from consumers who were confused by the great range of organic labels of farmers' marketing organizations (Hamm and Hummel, 1988) and had doubts about whether the organic products were truly organic. Consequently, in the late 1980s some supermarket chains stopped their sales of organic products (e.g. in the UK) or reduced their organic offerings (e.g. in Germany).

In the early 1990s, market structures were influenced a lot by politics. The EU introduced support schemes for organic farming, which led to a rapidly growing supply, and the EU took responsibility for organic labelling so that consumers could be assured that organically labelled products were really organic. Similar regulations were introduced in non-EU countries of Europe, such as Austria (non-EU at that time) and Switzerland. Thus, the organic market became more attractive for supermarket chains through governmental guarantees of organic origin and through higher sales volumes. While demand for organic products steadily grew with several crises in conventional agriculture (see Chapter 8, this volume), supermarket chains became the most important sales channel for organic products in many European countries around the turn of the millennium. Conventional supermarket chains took the lead in the quickly growing organic market as they offered organic products at lower prices because of significantly lower distribution costs compared with organic food shops or health food stores. Another main advantage of conventional supermarkets was the 'one-stop-shopping' they offered to consumers who were not willing to make extra trips to buy organic foods in special stores. Finally, consumers' doubts about the organic origin of the products offered in supermarkets decreased because of the EU regulation on organic products.

Interestingly, the story is nearly the same all over the world. After conventional supermarkets started offering organic products, sales were boosted in North and South America, Asia and Oceania. The crucial role of conventional supermarket chains can easily be explained by their ability to reach the majority of consumers and meet the needs of their buying behaviour. An interesting phenomenon is that in most countries of the world, traditional channels such as direct sales by farmers or sales through natural or health food stores also increased after conventional supermarket chains started selling organic products. Normally, a big company's involvement in organic food is accompanied by advertising and public relations campaigns so that the public becomes more interested in organic food in general. Since the product range of conventional supermarket chains is usually limited at first, consumers interested in trying several different organic products must look for them in specialized shops. Thus, conventional supermarket chains entering the organic market not only lead to more competition regarding price and quality of organic products, but also create sales opportunities for traditional organic food suppliers in the form of new organic customers seeking regional organic food specialities that are not offered by big chains.

Since the late 1990s, retail chains offering a wide range of organic foods have become very successful in many countries. The most prominent example is the US chain Whole Foods Market, with a turnover of several billion US dollars in the US market and with shops in the UK too. In Europe, such specialized supermarket chains exist mainly in

Germany and the UK, but on a much smaller scale. Big conventional supermarket chains, such as Albert Heijn in the Netherlands or Rewe in Germany, also started specialized organic supermarket chains. Despite the similarities of some developments in the two leading markets of the world, one big difference between organic supermarket chains in the USA and Europe is still that while specialized retail chains in Europe offer a wide range consisting exclusively of organic products, specialized US chains such as Whole Foods Market offer many other 'healthful' products in addition.

With these changes in the market, consumers' demands also changed greatly. While in early stages of the organic market, consumers and traders accepted a lot of imperfections, they became more demanding toward their suppliers as the markets grew. Big market structures demand large quantities of products of consistent quality, and thus a coordination of production. An organic offer side-by-side with comparable products from conventional agriculture in conventional supermarket chains implies comparable product quality regarding appearance and freshness. This is not easy to achieve, as most conventional treatments to keep the products looking fresh are not allowed in organic farming. Furthermore, today's consumers of organic products demand the same wide product range in organic quality as is offered for conventional products. This includes all kinds of fresh tropical and out-of-season products as well as the full range of processing, from fresh to frozen, from natural to low fat or enriched with vitamins or minerals. Another widely accepted requirement is the combination of organic production with fair trade, especially for organic products imported from developing countries. More and more organic products from developing countries therefore carry both an organic and a fair trade label in Europe and the USA.

In countries with high market shares for organic products, the differences left between organic and conventional foods became smaller and smaller, besides the differences between their respective agricultural production systems. Therefore, some consumers and some suppliers became dissatisfied because organic stopped being something very special. Consequently, suppliers looked for new opportunities to differentiate their products from the mainstream organic market through additional ethical values (produced by mentally handicapped persons, local, animal welfare) or additional services (pick-your-own vegetables, rent a layer hen, vegetable box schemes). Therefore, in countries with high market shares for organic food an increasing differentiation among organic food products can be observed. Discount strategies as well as premium price strategies are followed by different (and sometimes the same) suppliers under different labels or in different shops. In countries in which organic products are quite new and still very special, differentiation among organic products is not an important interest; rather, the main task is still to communicate the differences between organically and conventionally produced food.

7.4 Institutional Changes in the Market

7.4.1 New policies for market development

As mentioned earlier, changes in policies had major impacts on the organic market. In the following, we describe first the standard setting and second the policies that have played an important role in the development of the organic sector.

In 1980, IFOAM adopted the first version of its private Basic Standards. They were the first internationally agreed-upon standards for organic farming, and fundamentally influenced the later development of the governmental standards of the EU in 1991 and 1999 and the joint standards of the FAO and WHO Codex Alimentarius Commission in 1999 and 2001 (Geier, 1997; see also Chapter 8, this volume). Since the latter are considered a reference for decisions of the WTO, they will play an important role in future international organic trade should trade disputes arise (De Haen, 1999).

The EU-wide governmental organic standard EC Regulation 2092/91 (in 1991 for plant products and 1999 for animal products) established uniform rules for organic producers throughout Europe, protected the market from fraud through safeguarding the term 'organic' and provided higher assurances for consumers as well (Dabbert et al., 2002). Jointly with the agri-environmental programmes, it gave a boost to the organic market in Europe (Michelsen et al., 1999). It made trade easier between European countries and also with countries outside the EU because they had to comply with one common regulation. Since then, EC Regulation 2092/91 has been a model for many national regulations in countries outside the EU, which are obliged to comply with it to be able to export to the EU anyway. At the same time, however, it imposes high standards and costs on an infant industry and therefore potentially hampers their internal markets. Countries that have gained easier market access to the EU by being accepted on the 'third country list' on the grounds of having equivalent national regulations are Australia, Argentina, Costa Rica, Israel, New Zealand and Switzerland.

In the USA and Japan, the two other major markets besides the EU, national standards and regulations came into force in 2002 and 2001, respectively. In both countries private market actors had largely favoured the development of national regulations because they wished to be protected from fraud. However, the first draft organic standards issued by the US Department of Agriculture (USDA) brought forth strong objections because they did not ban the use of genetically modified organisms (GMOs) (Vaupel, 1999), and thus offered a notable example of a government potentially influencing the organic market. (Because of a massive public outcry, this proposal was withdrawn, with the final regulation totally prohibiting the use of GMOs). With the implementation of the National Organic Program (NOP) in 2002, only organic products meeting the USDA's standards are allowed to be marketed. The NOP and the strong USDA label have given the industry a boost through creating higher consumer awareness. The Japanese Agricultural Standards restrict the market to products certified by an accredited organization. This caused products not complying with this requirement to lose their status, temporarily reducing the market volume to one-tenth (Willer and Yussefi, 2004). Future facilitation of international organic trade depends on the mutual recognition of the regulations in the major markets and producing countries.

The first support schemes for organic farming in Europe were implemented in the late 1980s. Denmark led the way with a groundbreaking law on organic farming in 1987. It was the first national programme that defined organic farming and supported its expansion (Lampkin et al., 1999). In 1988, the EU implemented the so-called extensification scheme (EC Reg. 4115/88), whose main objective was to reduce surpluses of conventional agricultural products. Germany was the first country to introduce a scheme to support conversion to organic farming financially in the context of this policy, with the result that the organic area increased ninefold in only 4 years (Hamm and Michelsen, 1996) (a detailed account of policies for organic farming is given in Chapter 6, this volume).

An even bigger boost was provided by EC Regulation 2078/92. This regulation consisted of an agri-environmental programme considered as 'accompanying measures' for the major reform of the EU's Common Agricultural Policy in 1992, known as the McSharry Reform. The measures were implemented differently in different member countries. Because of that, the organic sector took different paths of development in the various countries. For example, the lag in the development of organic production in the UK was partly due to the lack of financial support for already-converted farmers. The supply in the organic market increased notably as a result of the financial support established throughout Europe from 1994 on (Lampkin et al., 1999; Dabbert et al., 2002). The subsidies were more attractive for extensive land (grassland, olive groves and vineyards) in disadvantaged areas than for arable land with good soil fertility. This increasingly led to an oversupply in organic beef, milk, lamb, olive oil and wine, while organic poultry, pork and vegetables were in short supply (Hamm *et al.*, 2002; Hamm and Gronefeld, 2004).

Over the years, the oversupply in some organic product groups has become a serious threat to the organic market. The government of Denmark, therefore, was the first to change its policy to one that involved more demand-strengthening, with its national action plans of 1995 and 1999 (Dabbert *et al.*, 2004) supporting the organic sector not simply through production subsidies, but rather taking a more integrated approach. By the late 1990s other European governments followed with their own national organic action plans (Dabbert *et al.*, 2004). Several European governments have recently initiated a change in support of organic farming on the European level. Collective Measures of Support were decided in 2004 in a European Action Plan for Organic Food and Farming, such as harmonization of national rules and improvement of consumer awareness, data availability and monitoring systems (European Commission, 2004).

As an overall development in organic standard setting and policy, two trends must be noted. First, governments increasingly have taken over decision making on standards from the organic movements. This has had both positive aspects – enforceable laws, safeguarding the market framework – as well negative ones, e.g. the previously mentioned example of the US proposal (subsequently withdrawn) that would have allowed GMOs in organic production, thus threatening the trust of consumers and going against the will of the organic movement (Vaupel, 1999; Sligh and Christman, 2003). Second, the financial support led to a high dependence of organic producers – and as a result the whole organic market – on governmental decisions and budgets. The reason for this development is that when production is subsidized, the burden of the extra cost of organic production is partly taken over by the government, and market principles lead to a decrease in the organic price premium for the producer.

7.5 Structural Developments in the Supply Chain

Over the years, the organic supply chain has undergone major changes. Three interlinked trends may be discerned.

First, the development of the organic market was and is characterized by increased competition and professionalism. Oversupply and lower price premiums have tightened the market, and competitors have had to improve their performance and quality as market principles became effective. For this reason, actions that were done out of idealism cannot be continued if consumers do not pay for them, either because they are unwilling to do so or because the issues have not been properly explained to them. Broadening the range of organic products and meeting the special needs of both consumers and retailers through innovations such as ultrapasteurized milk have also become more important for staying competitive. Smaller producers or processors especially are under increasing economic strain and potentially excluded because they cannot profit from economies of scale and because the retailers and supermarkets need bigger deliveries. Also, the competition for shelf space favours larger and financially flexible rivals (Sligh and Christman, 2003).

Second, pioneering organic companies have been growing and the market has undergone concentration. These companies have been successful, but needed to grow to survive in the increasingly competitive market. Examples are the German processor and distributor Rapunzel, founded in 1974 and now employing 240 people, and Britain's first organic dairy, Rachel's Organic, which now accounts for around 10% of the British market (Rachel's Organic, 2004; Rapunzel, 2004). The distributor Tree of Life, founded in 1970 in the USA, is the largest organic food distributor worldwide, with annual sales of US$3.5 billion (Sligh and Christman, 2003). The increasing size of the companies and the requirements of the market demanded modifications of the legal form and organizational structure, and with the change of generation, ownership shifted into new hands, in some cases causing drastic changes in the company's self-conception. Increasing professionalism and expanding companies also lead to market concentration. For example, Horizon Organic Dairy processes 70% of the organic milk in the USA (Sligh and Christman, 2003).

Finally, large conventional processors, distributors and retailers increasingly have become involved in the organic market. Seeing the organic sector gaining market share, with annual percentage growth rates in the two-digit range and organic products entering supermarkets, such companies became interested in investing in their own organic product lines or acquiring organic companies. Global players such as Coca-Cola, Groupe Danone, Heinz, General Mills, Nestlé and McDonalds are taking part in the organic market today. Groupe Danone bought into the US organic yoghurt manufacturer Stoneyfield Farm, while Heinz owns part of the Hain-Celestial group, the world's largest processor of natural and organic foods, and McDonalds soon hopes to sell 3.4% of British organic liquid milk through its fast food restaurants. The previously mentioned pioneering distributor Tree of Life is now owned by a Dutch corporate group of food companies.

On the retail level, major national conventional retailers account for large fractions of national organic sales, e.g. Albert Heijn (around 45% of the Netherlands' total), and Tesco and Sainsbury's (each 30% of the UK total) (Sligh and Christman, 2003). In Switzerland, around 80%

of organic products are sold through the retail chains Coop and Migros (Richter, 2004). Creating their own organic brand gives retailers greater market power, with interchangeable producers and processors in the background. The ownership and decision-making power is increasingly taken out of the hands of the pioneering and all-organic market actors, resulting in a growing but changed organic market.

7.6 The Changing Organic Food Consumer

The emergence of biodynamic agriculture in the 1920s and other forms of organic farming in the 1930s and 1940s was followed by the creation of an organic farming movement whose members were involved in a number of practices that included eating organic food (Lockie et al., 2001a). Very little information exists on those early consumers. In 1946, inspired by the writings of Eve Balfour, the Soil Association was founded in England by a group of farmers, scientists and nutritionists who believed there was a direct connection between farming practices and the health of plants, animals, humans and the environment (see Chapter 10, this volume).

The Reformhaus movement in Germany also started in the years following World War I, mainly connected with a renewed interest in natural healing systems ranging from herbal remedies to diet and nutrition (as discussed in detail in Chapter 2, this volume). The term *Reformhaus* nowadays means a health food shop, and is often taken as a synonym of *Naturkost* (natural shop) and *Bioladen* (organic shop), but that was not so at first.

Since the 1970s, health food shops in Europe, the USA, Australia and Japan have increased the variety and quantity of organic foods they sell, and specialized organic shops and restaurants have increased in number, further expanding the availability of organic food to consumers (Clunies-Ross, 1990; Belasco, 1993). In the 1980s and 1990s, mass merchandisers such as supermarkets and hypermarkets further increased the number of consumers potentially exposed to organic food, thereby increasing the number of consumers further and changing the patterns of consumption.

However, consumer demand for organic food has been extensively studied only since the late 1980s, because of growing consumer interest in these products. Organic foods' share of total food consumption is still below 1% in most of the world, if we exclude a small group of countries in northern and western Europe, e.g. Austria, Germany, Denmark, Sweden and the Netherlands (Hamm and Gronefeld, 2004).

Unfortunately, systematic official statistics on consumption are still not available, and studies of consumers' purchasing behaviour often rely on self-reported data. Many private market-research companies are collecting point-of-sale data on organic food consumption patterns, but most of these studies are not generally available to the public. However, there have been several studies in various countries on consumer attitudes and preferences with respect to organic food.

As mentioned earlier, in most developed countries consumer interest in organic food has increased during the past decade or two. The growth in organic consumption can only partially be ascribed to an increasing interest in the quality and safety of food products. Food scares and scandals in Europe at the end of the 20th century influenced consumer perceptions of organic food, but they did not represent an enduring factor that explained the boom of organic food consumption.

Although health and safety issues are quite relevant, organic market development depends on the outcome of several evolving issues influencing both the demand and the supply side (Torjusen *et al.*, 2004). As we have seen in the previous section, growth in organic demand can be ascribed mainly to 'external' factors (Hill and Linchehaun, 2002): the increased availability of a wide range of organic foods in large conventional retail channels coupled to a higher consumer product awareness because of the launch of highly promoted logos (e.g. the BIO-Siegel in Germany).

7.7 What We Do and Do Not Know About the Organic Consumer

A wide range of consumer studies have tried to understand the underlying reasons for purchasing organic food. However, in order to understand consumers' motivations, discussed in detail below, it is first necessary to understand who the organic consumer was and is.

7.7.1 Demographics

In early studies, the 'typical' consumer of organic products was initially identified as a female between 30 and 60 years, living in a city or a large town, with an average or above average education, and in the upper middle or upper income bracket. However, more recent studies have shown that age, gender and income no longer should be considered as distinguishing characteristics of organic consumers.

The age of typical organic consumers, which has been reported differently across countries (18–49 in the USA, 30–40 in Japan, 25–50 in the Netherlands), is mainly correlated with cultural differences and the consumers' level of involvement in health and environmental issues

(Lohr, 2000). Age and gender of purchasers have varied over the years, making any projections unreliable. Furthermore, according to many recent studies, women were usually overestimated because of sample bias: among consumers, fewer men are responsible for household purchases (Beckmann et al., 2000; Lohr, 2000). High product prices often seem to be the most relevant barrier to the growth of organic demand in both high- and low-income countries; the high price premiums of many organic products could explain why even some high-income consumers are often not attracted to them, but the significance of price premiums is declining.

Nevertheless, at the moment, regular consumers can commonly be found among people living in urban areas and with higher educational levels (Torjusen et al., 2004), and often with young children. However, this consumer profile is changing with the increasing number of shops offering organic foods. This change can be ascribed to three factors explored at greater length in Sections 7.8 and 7.11. First, income no longer seems to be such a discriminating factor in organic consumption. Because of lower premiums and a higher perception of the quality of organic foods, many more consumers are willing to pay more for organic quality. Second, the higher availability of organic foods in the mainstream retail sector has increased consumer awareness of organic foods and encouraged many more consumers to try them. Third, organic consumption has become more of an attitude rather than just a matter of reasoned choice because of new lifestyles and changing consumer preferences (Zanoli, 2004).

As a result, organic consumers no longer represent a restricted population group defined by particular characteristics. In the USA, not only whites, but also African Americans, Hispanics and Asians buy organics (Hartman Group, 2004). In Europe, the organic consumer profile is broadening too. In some countries, such as Germany, Denmark and Switzerland, where a large majority of consumers have tried organic food at least once, there no longer is much point in studying the sociodemographic characteristics of the occasional organic consumer, since they probably match those of the general population (Beckmann et al., 2000).

7.7.2 Knowledge and awareness

The level of information about organic products is still very variable both within and between different countries and consumer groups. Overestimation of consumer knowledge, because of unstandardized protocols for data collection, complicates the correct interpretation of organic market data. Several studies in Europe addressed the knowledge issue. Although they showed that high percentages of the population knew about 'organic food', they did not check precisely what consumers understand by the term; this often leads to erroneous and inconsistent results. In the past, it was taken for granted that consumers knew what organic products were and were sufficiently informed about them. These surveys relied on quite high knowledge levels. But when other studies checked actual consumer knowledge by asking what consumers really meant by these products and what they knew about them, the results showed more limited knowledge. Recent consumer studies (Torjusen *et al.*, 2004) reveal that when the procedures used to collect the data are correctly chosen, consumers' knowledge is still low. Occasional consumers, especially, are still confused about products labelled as 'natural', 'low input', 'integrated' and 'organic'. They also show little interest in getting more information and in many cases do not know anything about regulatory standards.

Across Europe, a recent survey confirmed these findings, although indicating that in general terms, consumers' awareness of organic food has increased: only a minority of consumers are unaware that organic food exists (Zanoli, 2004). The survey also found that greater knowledge about organic products is associated with higher consumption and higher education level, but not with gender or income. It also showed that consumers' knowledge about organic farming and organic certification is still very low, but also varies greatly among countries. In countries such as Austria, Finland, the Netherlands, Spain and Denmark, consumers show that they know something about organic farming, regulations and standards of organic processing, but most have just a vague idea of the organic concept. In others, consumers have recognized better the relationship between organic production and the reduced use of pesticides and chemical fertilizers.

7.7.3 Frequency of consumption

According to national investigations, the frequency of organic purchases, despite wide variations among countries, has generally increased. The figures are not strictly comparable across countries because estimates rely on unofficial, unstandardized market research that varies in sample size, investigation methods, etc. This situation is even worse in the Pacific region (Australia, China, Japan and Taiwan) and North America (Canada, Mexico and the USA), where the delay in establishing accepted organic national standards in some cases has even caused overestimation of the organic demand (Lohr, 2000). In these countries, where consumer awareness of organics is still lower than in Europe, surveys are probably leading to misrepresentation because of respondents' different understanding of 'organic'.

Nevertheless, bearing in mind these caveats, among non-European countries Canadians show the highest consumption frequency. Most (71%) have tried organic food at least once, and about 18% identify themselves as regular buyers (Environics International, 2001). In Japan, half of organic buyers are 'regular' consumers (International Trade Centre, 1999), but the number (and percentages) of organic buyers within the population has not been investigated. In Australia, 40% of consumers claimed they consumed at least some organic food, and 6.7% ate organic foods exclusively (Lockie *et al.*, 2001b). In recent years, Australians have increased usage of organic food (in terms of higher frequencies of use) by over 20%, with dairy and cereal products the most important categories, while the percentage of non-organic users has decreased. The figures are lower in Latin America, e.g. 11% are relatively high users in Mexico and only 6% in Brazil (Health Focus International, 2003). These two countries, according to surveys, have a high potential for the domestic market, but lack supply and purchasing power.

In Europe, the highest percentages of regular organic consumers are in the UK and Germany (up to 65% of all consumers regularly buy one or more organic products), followed by Austria (two-thirds buy occasionally and 20% are regular consumers) and Switzerland. The lowest percentages are in Italy (4%) and the Netherlands (5%) (Lohr, 2000). On the other hand, in recent years the number of regular buyers in Europe has not increased much, although according to some surveys, a large part of the most recent market growth has been due to an increase in per capita spending by regular users, with much less due to the increase of occasional purchases by new consumers (ISMEA, 2004).

7.7.4 Product categories

Consumer surveys reveal that fruits and vegetables are the leading categories for organic sales. These consistently are purchased most often and are the categories first purchased by non-regular organic consumers. In the future, since organic food is expected to expand beyond its traditional base, many more mainstream grocery groups are going to enter the organic sector to satisfy consumer needs. Meat and cereals have a great deal of potential growth, since organic shoppers are looking for the same convenience and range of food they find in the conventional offerings. Consumers also ask for organic snacks, prepared meals (including frozen food) and beverages. Organic baby foods are another 'hot' category (Canadian Grocer, 2003; Health Focus International, 2003; ISMEA, 2004; Organic Monitor, 2005).

7.8 Consumers' Motives for Purchasing Organic Food

Although most research has generally stressed consumers' positive attitude toward organic products, it is still difficult to explain the low level of consumption of these products. Across countries, using both qualitative and quantitative methods, studies trying to assess the reasons behind the purchase or rejection of organic products indicate several factors driving organic consumption.

Most studies undertaken throughout the world identified health, taste and environmental concerns as among the most important motivations for organic purchases (Hartman Group, 2004; Whole Foods Market, 2004; Zanoli, 2004). In some cases ethical issues have emerged among consumers: animal welfare and social motives (like supporting local farmers/community) are other reasons for choosing organic products. The reasons for purchasing organic food are similar across countries.

Health, which was the dominant factor for earlier consumers, is still mentioned as the most important motive behind organic purchases. Nevertheless, since health concerns alone do not seem to have a strong impact on demand, recent studies have tried to go beyond this multifaceted motive. The health aspect usually emerges in two forms. First, in most cases it involves a reduction in health-related risks, such as by avoiding pesticides, synthetic additives, hormones, antibiotic residues, GMOs and other threats such as BSE. However, some consumers expressly mention positive nutritive aspects and the possibility of improving health by eating organic food (Zanoli, 2004). In both cases, although high percentages of consumers believe that organic food is more healthy than conventional food, few actually buy it.

Very recent qualitative studies surveying both European and non-European consumers and going deeper into consumer perceptions of organic food (Brunsø et al., 2002; Zanoli et al., 2003; Zanoli, 2004) have shown that organic food consumption is a lifestyle choice, often related to important occurrences in life (e.g. pregnancy, birth of a child, ageing, disease). These studies seem to confirm the importance of the health factor and its closeness to food's naturalness. They also reveal that pursuing health by buying organic foods is no longer a simple means of avoiding health problems (Health Focus International, 2003), but is mainly correlated with a good physical and emotional feeling together with a wide-ranging sense of well-being and quality of life (personal end goals). Besides health, also mentioned among the most important motivating factors are animal welfare, food as enjoyment, environmental concerns and trust (Zanoli, 2004). In northern Europe, for example, concern for animal welfare seems to be especially important to British consumers, whereas Swiss consumers place more importance on good husbandry conditions than on personal health when thinking of meat consumption.

In general, altruistic values such as environmental protection, respect for nature and animal welfare are not as strong as personal health and the pleasure of eating good food.

Sensory appeal, which according to Australian consumers had no relevant significance (Pearson, 2001), is an increasingly important factor for Canadian and American organic consumers. In Europe, the pleasure of eating good food has also grown as an influence in choosing organic food (Zanoli et al., 2003; Zanoli, 2004). Regular organic consumers usually have a more positive perception of the aesthetic, sensory and organoleptic characteristics of organic than conventionally produced food. On the other hand, most occasional consumers also mention bad taste as a negative attribute that prevents them from buying organic food. Altogether it appears that consumers are less willing than in the past to forgo taste in favour of health benefits, and show a certain desire to avoid compromises because improving their quality of life is among their main priorities.

7.9 The Importance of Certification and Labels in Building Trust

Organic marks and labels are used in different ways and have different roles in various countries. Their credibility is just partially related to the history of the brand (or label), and mainly depends on trust relationships that consumers perceive. Since labels and certifications have a short history inside the organic sector (apart from Denmark, no other country seems to gain benefits from a well-known and established organic label), consumers have to rely on trust attributed to other persons (e.g. other consumers, shop attendants) or institutions (e.g. their Ministry of Health).

Organic labelling is particularly important in mainstream retail channels, where the lack of direct personal contact with producers or skilled staff creates a gap between the consumer and the product. In supermarkets, consumers need to differentiate between organic and conventional foods in a simple and quick way. Organic labelling enables this distinction to be made, thereby ensuring authenticity and allowing consumers to choose the organic quality attributes (Midmore et al., 2005).

Consumers need to be trustful because the 'organic' characteristics of the product are not visible, but according to the most recent EU survey (Zanoli, 2004), consumers seem quite confused in choosing which trust-building factor is most reliable to them (labels, certification bodies or farmers). To trust the healthfulness of organic food, people want to be sure that both producers and certification bodies are reliable and follow the rules. However, in many cases consumers say that regulations are not so easy to understand; they do not know much about them, but nevertheless often declare that the standards are not rigorous enough. These arguments get stronger for occasional consumers who are less keen to pay higher prices for organic food. Consumers with less expertise and less involvement, who show less use of intrinsic cues (sensory appeal and appearance) to judge quality, also have more need for quality signs (Brunsø et al., 2002).

Most European consumers have a positive attitude toward existing labels (Naspetti and Zanoli, 2005), but unfortunately, most usually know and trust only a few labels, some private and some public. Some countries (e.g. Denmark) have benefited from having a unique national organic label that is highly appreciated by consumers. Others, such as Germany, have not particularly benefited from the introduction of the national public organic logo (BIO-Siegel). Its impact on consumption has not been so important, while other factors have depressed organic demand in Germany, including a bad economic cycle and the Nitrofen scandal of 2002, involving a banned herbicide found in organic feed.

The large number of certification bodies and different labels that appear in the market contribute to confusion and lack of trust for many European consumers. Some consumers are not keen to trust organic labels at all. In Italy, Austria, the UK, Switzerland and France, consumers often have more confidence in farmers, retailers and public institutions than in any quality sign (Zanoli, 2004). However, because direct sales from farms account for only a small portion of sales in Europe, a personal relationship cannot be the key to higher consumer confidence in organic products, unless it is coupled with other means that can be applied in large retail channels, such as traceability systems. Indeed, many organic producers and retailers have started to use electronic traceability codes (to be used on an Internet site to learn the name of the producer) or other systems (Tesco places a picture and some information about the farmer on some of its products) to enhance consumer trust in organic products. Consumers feel more reassured by the potential possibility of a personal check of the production by the farmer. Unfortunately, this is not always easy, especially with multi-ingredient and highly processed products, the ones that are less trusted by organic consumers.

Consumers are asking for more information, stricter controls, higher transparency and 'clear' (i.e. easy to understand) standards. Unfortunately, standards are bound to get more complicated (as discussed at length in Chapter 8, this volume), especially for processed products. As a result, despite a wide variation in certification programmes and regulations in Europe, trust in organic labels is strongly connected to a need for more information and transparency.

The labelling issue is directly related to product knowledge. Consumers who know less about organic food are also the ones who are not fully confident in organic standards and inspection schemes. To fill the knowledge gap, these consumers request a quality sign that gives information about the certification body, about how seriously inspections are conducted and about product standards.

Some consumers are asking for many other types of information to be placed on the label (information about the producer, about the product's place of production and origin, and about the product itself and the way it is produced). The issue of product traceability was already mentioned. The need to know the origin of the product is also connected to the desire that the products be produced locally or at least domestically. Again, in consumers' minds, this means that one can personally check on the farmer or the producer.

7.10 The Importance of Origin

Various surveys have shown that among European consumers, local origin evokes a positive feeling: it looks 'friendly' and is usually associated with small-scale production (Zanoli, 2004). Moreover, worldwide, lack of consistent standards affects the quality perception of overseas organic products.

Most regular organic consumers seem to prefer locally grown organic food, though the term 'local' is not always so clear and definable. Sometimes it is intended as a synonym of 'domestically produced', but usually it also refers to production that is very close or within the region. However, the appropriate distance that an organic food product should travel is not clear. There is clearly a contradiction between consumers' statements and their actual choices in the market. The organic fruit most requested in Europe is bananas, which are also the most traded worldwide (International Trade Centre, 2001); clearly they are not locally grown in Europe.

There are three reasons behind a preference for local organic foods: environmental (connected with the 'food miles' issue and ecological footprint); product quality (consumers think that local origin is a guarantee of quality characteristics such as naturalness, authenticity and freshness); and ethical (some consumers like the idea of supporting local farmers and producers).

Consumers' interest in locally produced organic food, as well as in the origin labels, is mainly due to the increased distance from production to consumption (Torjusen *et al.*, 2004). As discussed by Vergunst (2001), local food systems replace impersonal exchange with personal relationships of trust. Trust in local food systems, as a known social structure, is enhanced by the perception of a possible personal contact with the producer for evaluating the quality of the product.

In Europe, local preference is strongest in Austria, Germany, Switzerland and the UK (Zanoli, 2004). In some cases local products are even a substitute for organic products, and can compete with them. In the USA, Japan and Sweden, local products are preferred over organic products (Lohr, 2000). In Australia, according to a recent study, consumers were concerned about not being able to purchase local products once they were looking for organic products (Turnbull, 2000).

This is also true for occasional consumers, especially those living in the countryside (Zanoli, 2004). They often say that they are not interested in buying products with organic quality since they have their own home-grown vegetables and fruits, which they trust more. These consumers seem more sceptical toward organic produce and less interested in food certification, often because of lack of information.

7.11 Barriers to Further Expansion of the Organic Market

Price, availability and place of purchase are considered key factors encouraging or discouraging conventional food buyers and occasional organic consumers to buy organic food (Pearson, 2001; Hill and Linchehaun, 2002; Zanoli, 2004). Higher prices are usually stressed by non-organic consumers as the main reason for not purchasing organic food. In reality, the very first reason that consumers do not buy organic products seems mainly related to lack of information about them. In most cases, consumers say, they had never really considered them before (Howie, 2004; Zanoli, 2004).

7.11.1 Price

With expansion of the market, the price of organic food has become more competitive, but remains an important barrier for occasional organic consumers and non-users worldwide. However, price becomes less important among regular consumers when their experience increases and organic food becomes a lifestyle choice. Surprisingly, regular consumers complain more about the higher price since organic food represents a higher share of their food budget (Zanoli and Naspetti, 2002). However, once consumers have tried organic products they value them more. Price perception is influenced by various factors, but income does not seem to substantially influence consumers' willingness to pay. Organic food is expensive, but prices are not always that much higher if compared with high quality (premium) conventional food brands. Therefore, consumers' failure to recognize organic foods' additional value, not just their limited household budgets, makes them less willing to pay higher prices (Zanoli, 2004). Furthermore, value for money is influenced by both the type of consumer and the product category. For example, mothers attribute a higher value to children's food than to other organic product categories.

7.11.2 Availability

Most consumers surveyed worldwide declare they would be willing to eat organic foods more often if a larger variety was readily available in the market. Lack of availability was the most relevant barrier for the development of an organic market and the involvement of the mainstream retail channels. Even today there still is a perceived low availability within the store (Howie, 2004).

Two aspects of availability are relevant for consumers: absolute and relative availability. Absolute availability was the original problem. The organic supply was insufficient in quantity and variety to fulfil a wide consumer demand, and therefore organic products were found only in niche markets. Only when the organic supply was increased (often pushed by agricultural policy) did large retailers find that they could offer organic foods.

Relative availability is subtler. Even if organic products are in greater supply, consumers often cannot find them in their customary shops. Not all retail chains have committed themselves to stocking organic products, and some still carry only a few long-storage products. Besides, organic products are often not available in as full a range as conventional products. This frustrates consumers, who then tend to become less loyal to the organic idea.

Lack of availability in shops where people are used to buying is also related to the time and effort of shopping. The new consumer is constantly running and hates wasting time. The perception that finding organic products can be inconvenient, laborious and time consuming discourages less-experienced consumers. They are also reluctant to use organic foods because they fear that they do not know how to prepare them.

7.11.3 Place of purchase

As discussed earlier, most organic products were originally sold in natural food stores and farmer cooperatives. These channels are now dwindling in importance for both occasional and regular consumers.

In contrast, more and more supermarkets are selling natural and organic foods, and the principal place to purchase organic food is the supermarket. In the USA, about four in ten consumers indicated that it is at least somewhat important for their primary grocery store to offer natural and organic products (Food Marketing Institute, 2004).

In Europe, too, supermarkets are the preferred point of purchase for organics (Hamm *et al.*, 2002). An increasing array of organic products is available at more locations, which means better accessibility, convenience and easier shopping. Consumers select these stores to buy organic foods for practical reasons related to a desire to save time, and because they trust them. However, some consumers heartily reject them for a related reason. Despite increasing interest in bigger outlets, in most European countries, consumers show a remarkable level of mistrust toward these points of sales when organic purchases are concerned. Consumers' mistrust of large retail outlets ranges from 18% in Germany to 67% in the UK. Consumers seem to reject supporting mass-marketing structures because of a negative image resulting from poor quality products, leading in turn to health concerns, and from the unpleasant atmosphere they have experienced. Organic consumers complain about supermarkets as being crowded, chaotic and noisy, and discount shops being untidy and dirty. Their appearance reduces consumer confidence and trust. Given shoppers' uneasiness about the safety of food, the shopping place is very important when buying organic products. Trusting the quality of the products depends on overcoming two problems: doubts about the honesty of supermarkets that sell unknown brands that may not respect organic standards, and the risk of buying false organic products. Discounters are also criticized for their poor product presentation. In any case, supermarkets in most countries represent the preferred point of purchase for organic products, despite these declarations of mistrust.

Consumer interest in purchasing organic food in open-air markets or directly from farmers is not so important in Europe. Direct channels are preferred for providing good quality and fresher products, especially for fruits and vegetables, because contact with the producer builds trust. Lack of time seems to be the main barrier for shopping in these places. Small farmers markets are therefore very marginal as marketing channels. However, these markets seem to provide an occasion to try new products.

7.12 The Future of the Organic Market

A prominent issue emerging from current market trends is that despite its fast growth, organic production is still a niche market all around the world. Future growth will depend on many factors, but mostly on the new consumers' attitudes toward organic food.

Despite higher consumer awareness of organic food, product knowledge still appears low for both occasional and regular consumers. Researchers have found that there is still little knowledge of how organic products are produced and processed and do not know which characteristics are fundamental for consumers' perception of quality and safety. Improving the information level will not be enough. Primary producers, processors and other stakeholders in the organic supply chain have the difficult task of understanding consumers' complex, vague and sometimes contradictory requirements with regard to organic food quality. To understand these needs and to find out how to translate different conceptions of quality and food safety into practice, it is necessary to explore quality standards in much more depth. The need also emerges to resolve existing differences among different actors in the organic food chain and to determine how this can be achieved in a profitable way (Zanoli and Naspetti, 2006).

For example, people associate organic food with natural processes and with food products that are either unprocessed or have at most a low level of processing. However, modern lifestyles demand convenience products. Improving consumers' options for healthy products and natural production methods could help both to meet their needs and to reward producers' efforts. Because ordinary consumers will probably never become skilled evaluators of food, it is necessary to discover the simple indicators that they use to infer organic quality.

Consumers in some countries or regions have also become more interested in a local orientation and in the labels of origin of organic food because of the perceived increased distance between production and final consumption (Torjusen *et al.*, 2004). Further investigations should try to understand which safety and quality cues are used by consumers when buying local organic foods, how to overcome their mistrust, and how safety and quality issues could be better approached in an integrated 'farm to fork' approach to delivering product value.

There is also a need for a new positioning of organic products (Zanoli and Naspetti, 2006). Positioning based on well-being – eventually extended into a wellness concept embracing self-satisfaction and health – could be the way forward, since it would encompass all the core values that in cognitive terms represent the enduring appeal of organic food, and could trigger higher consumer involvement and loyalty.

Although market expansion will continue to depend on the stable core group of organic consumers, it will only be achieved with an increase in occasional organic consumers. However, this group appears to be more sensitive to price and convenience. Therefore, market expansion may rely on achieving economies of scale in distribution and greater levels of processing. The effect will be to shift the emphasis toward more profit-oriented supply chains. As noted in several places in this book, the desirability of such a shift will be very controversial.

References

Arbeitsgemeinschaft Landreform (1938) Anbau, Ablieferung und Verteilung biologischer Erzeugnisse. *Bebauet die Erde* 14(3), 32.

Beckmann, S., Brokmose, S. and Lind, R.L. (2000) Eco foods II – survey results, part 2 – Aspects of consumer behaviour. Working paper no. 5. Department of Marketing. Copenhagen Business School, Copenhagen, Denmark.

Belasco, W. (1993) *Appetite for Change*. Cornell University Press, Ithaca, New York.

Böckenhoff, E. and Hamm, U. (1983) Perspektiven des Marktes für alternativ erzeugte Nahrungsmittel. *Berichte über Landwirtschaft* 61, 345–381.

Brunsø, K., Fjord, T.A. and Grunert, K.G. (2002) Consumers' food choice and quality perception. Mapp. Working paper no. 77. Aarhus School of Business, Aarhus, Denmark.

Canadian Grocer (2003) Growing organically. *Canadian Grocer* 117(7), 52–58.

Clunies-Ross, T. (1990) Organic food: swimming against the tide. In: Marsden, T. and Little, J. (eds) *Political, Social and Economic Perspectives on the International Food System*. Avebury, Aldershot, UK, pp. 200–214.

Dabbert, S., Häring, A.M. and Zanoli, R. (2002) *Politik für den Öko-Landbau*. Ulmer GmbH & Co., Hohenheim, Germany.

Dabbert, S., Häring, A.M. and Zanoli, R. (2004) *Organic Farming: Policies and Prospects*. Zed Books, London.

De Haen, H. (1999) Producing and marketing quality organic products: opportunities and challenges. Paper presented at the 6th IFAOM Trade Conference, Florence, Italy. Available at: www.fao.org/organicag/frame2-e.htm

Demeter (2006) Das ist Demeter: Biologisch-dynamisch von Anfang an. Available at: www.demeter.de

Environics International (2001) Food issues monitor survey 2001. Available at: www.environics.net/eil/

European Commission (2004) European Action Plan for Organic Food and Farming. Brussels 10 June 2004. COM(2004)415 final. Commission of the European Communities, Brussels. Available at: europa.eu.int/comm/agriculture/qual/organic/plan/comm_en.pdf

Food Marketing Institute (2004) *Trends in the United States, Consumer Attitudes & the Supermarket*, Washington, DC.

Geier, B. (1997) Reflections on standards for organic agriculture. *Ecology and Farming* 15, 10–12. International Federation of Organic Agriculture Movements, Tholey-Theley, Germany.

Hamm, U. (1986) Absatzbedingungen bei Produkten aus alternativer Erzeugung. *Berichte über Landwirtschaft* 64(1), 74–152.

Hamm, U. and Gronefeld, F. (2004) *The European Market for Organic Food: Revised and Updated Analysis*. Organic Marketing Initiatives and Rural Development Series, Vol. 5. School of Business and Management, University of Wales, Aberystwyth, UK.

Hamm, U. and Hummel, K. (1988) Nahrungsmittel aus alternativem Landbau – Kleine Warenzeichenkunde. AID-Verbraucherdienst. Nr. 1218. 1st edn. AID Infodienst Verbraucherschutz, Ernährung, Landwirtschaft e.V., Bonn, Germany.

Hamm, U. and Michelsen, J. (1996) Organic agriculture in a market economy – perspectives from Germany and Denmark. In: Østergaard, T.V. (ed.) *Fundamentals of Organic Agriculture*, Vol. 1. Proceedings of the 11th IFOAM International Conference, 11–15 August 1996, Copenhagen. International Federation of Organic Agriculture Movements, Tholey-Theley, Germany, pp. 208–222.

Hamm, U., Gronefeld, F. and Halpin, D. (2002) *Analysis of the European Market for Organic Food*. Organic Marketing Initiatives and Rural Development Series, Vol. 1. University of Wales, Aberystwyth, UK.

Hartman Group (2004) *Organic Food and Beverage Trends 2004: Lifestyles, Language and Category Adoption*. Bellevue, Washington, DC.

Health Focus International (2003) Prairie-Organic conference trend survey. Presented at 'Prairie Wide Organic Conference 2004', 14–16 November 2004, Saskatoon, Saskatchewan, Canada.

Hill, H. and Linchehaun, F. (2002) Organic milk: attitudes and consumption patterns. *British Food Journal* 104, 526–542.

Howie, M. (2004) Industry study on why millions of Americans are buying organic foods. Organic Consumers Association. Available at: www.organicconsumers.org/organic/millions033004.cfm

International Trade Centre (1999) Organic food and beverages: world supply and major European markets. ITC/UNCTAD/WTO, Geneva.

International Trade Centre (2001) World markets for organic fruit and vegetables. ITC/CTA/FAO, Rome.

ISMEA (2004) Lo scenario economico dell'agricoltura biologica. Istituto di Servizi per il Mercato Agricolo Alimentare, Rome.

Lampkin, N.H., Foster, C., Padel, S. and Midmore, P. (1999) *The Policy and Regulatory Environment for Organic Farming in Europe. Organic Farming in Europe: Economics and Policy*, Vol. 1. University of Hohenheim, Hohenheim, Germany.

Lockie, S. Lyons, K. Lawrence, G. and Mummery, K. (2001a) Eating 'green': the relative importance of environmental concerns in the consumption of organic foods in Australia. Presented at the Kyoto Environmental Sociology Conference, 23–24 October 2001.

Lockie, S., Mummery, K., Lyons, K. and Lawrence, G. (2001b) Who buys organics, who doesn't, and why? Insights from a national survey of Australian consumers. Presented at the Inaugural National Organics Conference, 27–28 August 2001, Sydney.

Lohr, L. (2000) Factors affecting international demand and trade in organic food products. Faculty series, Department of Agricultural and Applied Economics, College of Agricultural and Environmental Sciences, University of Georgia, Athens, Georgia, USA.

Michelsen, J., Hamm, U., Wynen, E. and Roth, E. (1999) *The European Market for Organic Products: Growth and Development. Organic Farming in Europe: Economics and Policy*, Vol. 7. University of Hohenheim, Hohenheim, Germany.

Midmore, P., Naspetti, S., Sherwood, A.-M., Vairo, D., Wier, M. and Zanoli, R. (2005) Consumer attitudes to quality and safety of organic and low input foods: a review. Integrated Project No 506358, 'Improving quality and safety and reduction of cost in the European organic and "low input" food supply chains' (QLIF). Available at: http://www.qlif.org/research/sub1/pub/1_1_1_UWAL.pdf

Naspetti, S. and Zanoli, R. (2005) Consumers' knowledge of organic quality marks. Presented at the 15th IFOAM Organic World Congress, Researching Sustainable Systems. Adelaide, Australia, 21–23 September 2005.

Organic Monitor (2005) *The European market for organic fruit & vegetables.* London.

Pearson, D. (2001) How to increase organic food sales: results from research based on market segmentation and product attributes. *Agribusiness Review* 9, paper 8.

Rachel's Organic (2004) Organic pioneers for over 50 years. Available at: www.rachelsorganic.co.uk

Rapunzel (2004) Drei Freunde – Eine Vision. Available at: www.rapunzel.de

Richter, T. (2004) Mit Premium-Produkten Wechselkäufer gewinnen. *Ökologie und Landbau* 3, 17–19. Stiftung Ökologie und Landbau, Bad Dürkheim, Germany.

Sahota, A. (2006) The global market for organic food and drinks. Presented at Biofach Nuremberg 2006. Available at: orgprints.org/7255/01/sahota-2006-global-market.pdf

Sligh, M. and Christman, C. (2003) *Who Owns Organic? The Global Status, Prospects, and Challenges of a Changing Organic Market.* Rural Advancement Foundation International – USA, Pittsboro, North Carolina. Available at: www.rafiusa.org/pubs/Organic Report.pdf

Torjusen, H., Sangstad, L., O'Doherty, K.J. and Kjærnes, U. (2004) European consumers' conceptions of organic food: a review of available research. Professional report no. 4. SIFO National Institute for Consumer Research, Oslo.

Turnbull, G. (2000) Report on consumer behaviour in purchasing of organic food products in Australia. Master of Management thesis. Faculty of Business, University of Southern Queensland. Available at: dpi.qld.gov.au/extra/pdf/organicconsumers.pdf

Vaupel, S. (1999) Defining the standards. *Ecology and Farming* 21. International Federation of Organic Agriculture Movements, Tholey-Theley, Germany.

Vergunst, P. (2001) The embeddedness of local food systems. Mimeo, Dept. of Rural Development Studies, Swedish University of Agricultural Sciences, Uppsala, Sweden.

Whole Foods Market (2004) Trend Tracker Survey 2004, October. Austin, Texas.

Willer, H. and Yussefi, M. (eds) (2000) *Organic Agriculture Worldwide. Statistics and Future Prospects.* Stiftung Ökologie und Landbau, Bad Dürkheim, Germany.

Willer, H. and Yussefi, M. (eds) (2001) *Organic Agriculture Worldwide 2001.* Stiftung Ökologie und Landbau, Bad Dürkheim, Germany.

Willer, H. and Yussefi, M. (eds) (2004) *The World of Organic Agriculture 2004 – Statistics and Emerging Trends.* International Federation of Organic Agriculture Movements, Bonn, Germany.

Willer, H. and Yussefi, M. (eds) (2005) *The World of Organic Agriculture – Statistics and Emerging Trends 2005.* International Federation of Organic Agriculture Movements, Bonn, Germany.

Willer, H. and Yussefi, M. (eds) (2006) *The World of Organic Agriculture –*

Statistics and Emerging Trends 2006. International Federation of Organic Agriculture Movements, Bonn, Germany. Available at: www.orgprints. org/5161

Yussefi, M. and Willer, H. (eds) (2002) *Organic Agriculture Worldwide 2002.* Stiftung Ökologie und Landbau, Bad Dürkheim, Germany.

Yussefi, M. and Willer, H. (eds) (2003) *The World of Organic Agriculture 2003 – Statistics and Future Prospects.* International Federation of Organic Agriculture Movements, Tholey-Theley, Germany.

Zanoli, R. (ed.) (2004) *The European Consumer and Organic Food.* Organic Marketing Initiatives and Rural Development Series, Vol. 7. University of Wales, Aberystwyth, UK.

Zanoli, R. and Naspetti, S. (2002) Consumer motivations in the purchase of organic food: a means-end approach. *British Food Journal* 104, 643–653.

Zanoli, R. and Naspetti, S. (2006) The positioning of organic products: which way forward? Proceedings of Joint Organic Congress Organic Farming and European Rural Development, Odense, Denmark, 30–31 May, pp. 10–11.

Zanoli, R., Gambelli, D. and Naspetti, S. (2003) Il posizionamento dei prodotti tipici e biologici di origine italiana: un'analisi su 5 Paesi. *Rivista di Economia Agraria* 58, 477–510.

8 Development of Standards for Organic Farming

O. Schmid

Coordinator of EU Research Project ORGAP, Research Institute for Organic Farming (FiBL), Ackerstrasse, 5070 Frick, Switzerland

8.1 The Emergence of Standards

For a long time organic farming was generally understood as a more natural form of farming characterized mainly by the non-use of chemicals and other synthetic inputs. This rather narrow perception later changed when organic farming became defined in private standards and later in regulations, for example, by emphasizing a more preventive approach to crop and animal production, or by taking animal welfare issues or processing principles into account.

As long as organic farmers were selling their own products directly to consumers on the farm or in a market, there was no strong need for standards, inspection or certification. Consumers could always directly ask the farmers what they were doing and what inputs they were or were not using. But when the relationship between farmers and consumers changed, with the market becoming more impersonal, more centralized and more globalized, there was an evident need for standards and an inspection system both to protect the producer from unfair competition and to protect the consumer from fraud.

Another reason that standards were not very important in the 'pioneer' phase before the 1980s was that in several countries, organic farmers were often in very close and direct contact with the charismatic pioneers of the various movements. The few organic farmers at that time needed to cooperate closely with each other because they were often marginalized in the general farming community. This also resulted in a form of 'social control' among them. In Switzerland, by the 1950s, the founder of the organic-biological movement, Hans Müller, had established small regional farmer discussion groups, mostly with experienced farmers as group leaders, which he visited regularly and where the first ideas of organic farming were discussed (Fischer, 1982; Vogt, 1999; see also Chapter 2, this volume).

For several decades, organic farming was hardly recognized by the general public and consumers. One reason was that before about 1980 there was no commonly and broadly agreed definition of organic farming. When organic farming became a more political issue (a development described in Chapter 6, this volume), standards became more relevant, particularly when they were a condition for receiving direct support payments.

8.2 Private Pioneering Standards

The pioneer in setting standards was the biodynamic movement, which was initiated in Koberwitz in the 1920s by Rudolf Steiner. Koepf and von Plato (2001) report about the development of the biodynamic movement in their book, as does Vogt (1999) in his German thesis about the history of organic farming. Already by 1928 the first Demeter cooperative in Bad Saarow formulated very short private norms as part of the contract for farmers who wanted to use the name 'Demeter' for their products. Basically, these norms were reduced to three requirements:

1. Products must meet the same legal commercial standards (basic quality norms) as for non-organic products;
2. Seed must be from biodynamically cultivated fields where, in particular, no artificial fertilizers had been used for 3 years;
3. Arable land should have at least 3 years with no artificial fertilizers and should get biodynamically treated manure twice per year.

It is interesting that the concept of conversion period was already established at that time.

In 1931 the Demeter label was registered and was awarded to inspected farms. When the Demeterbund (an association for promoting Demeter products) was founded in 1955, a more formalized structure was created for the use of the Demeter label. A sharing of responsibilities for standards development was formulated in 1958 between the holder of the label on the one hand, and the institutions developing biodynamic agriculture on the other. These biodynamic standards, which were written more as guidelines, have continued to be further developed. In 1972, at a meeting of European Demeter movements, it was decided to develop international standards. This was the start of the international Demeter movement (Heinze, 1972), the same year that the International Federation of Organic Agriculture Movements (IFOAM) was founded (as described in detail in Chapter 9, this volume).

An approach similar to that of the Demeter movement was followed by the movement of Hans Müller in Switzerland, where for many years the first organic-biological cooperative Anbau- und Verwertungsgenossenschaft (AVG), founded in 1946, had a short set of norms as part of its delivery contract. Later, these norms were further developed by BIOFARM, another organic-biological cooperative in Switzerland (Scheidegger, 1993). They were the basis of the first Swiss standards of organic agriculture. Typical for these very first norms – better described as guidelines – was that organic farming was mainly characterized by the non-use of synthetic fertilizers and pesticides and especially by careful treatment of farmyard manure.

In the UK, by 1967 the Soil Association already published the 'First Soil Association Standards for Organically Produced Food' (see Chapter 10, this volume). These standards contained a hierarchy within each subsection with recommended, restricted and forbidden substances and practices (Soil Association, 1967). This hierarchy was later taken up by other standards (Jespersen, 1998).

In France, organic farming developed very early and was well documented (Aubert, 1970; De Silguy, 1991). In 1972, the organization Nature et Progrès had already started to establish a secretariat and a working group that developed a private standard (*règlement*). In 1974, the first 30 farmers were able to use a private logo for their products based on a private inspection system (Nature et Progrès, 1974). Later, in the 1970s, other organizations in Switzerland, Germany, Austria and the USA started to work out private national standards for their producers.

Already by 1976, a working group in Switzerland initiated by Hartmut Vogtmann started to work on a national umbrella standard for organic farming with minimum requirements. This was necessary because the state authorities wanted to forbid the term 'biological' by law (Scheidegger, 1993; see also Chapter 14, this volume). After intensive discussions involving several private label organizations, the first umbrella standard was published in 1980 on the basis of a common contract among five organizations (Scheidegger, 1993). This search for a common denominator among the standards for different versions of organic farming, including biodynamic, served as a model for the development of the IFOAM Basic Standards, which were developed at the same time.

8.3 The Role of IFOAM and its Basic Standards

In 1976, at the General Assembly of IFOAM in Seengen, Switzerland, it was decided that IFOAM should work on a common definition for organic farming. About 2 years later, the minutes at the 1978 IFOAM General Assembly in Montreal (IFOAM, 1978) reported that 'an urgent task was to give a more precise definition to what is meant by "biological", "organic", "ecological", and "natural" agriculture and to compile a common set of standards for produce sold under such a label or guarantee'. It is interesting that the issue of regional variation was already included in the task description, which mentioned that such a document should 'define the characteristics of biological (organic) agriculture from an evolutionary point of view, and taking into account the varying ecological circumstances obtaining in different areas'. A technical committee was established, chaired by Claude Aubert (France) and Otto Schmid (Switzerland).

In November 1979, the first draft of 'The basic rules of biological agriculture standards under consideration by IFOAM' was circulated among member organizations for comments. In 1980, the first version of 'Recommendations for international standards of biological agriculture' was accepted by the biannual IFOAM General Assembly in Brussels (IFOAM, 1980). In 1982, these standards became the 'Standards of biological agriculture for international trade and national standards', with validity restricted to 2 years (IFOAM, 1982).

8.4 The Role of Public and International Regulation

Several countries in Europe had already developed their own national regulations (e.g. France, Denmark and Austria) and logos (France and Denmark) for organic products in the 1980s, in some cases long before the European Union (EU) regulation on organic production came into force. These logos are well known today and are highly trusted by consumers, and are one reason for the boom in organic products in those countries.

The development of the worldwide market for organic products for several years has been increasingly determined by international and transnational governmental rules. The most important influences are EEC Reg. 2092/91 and the US Department of Agriculture's (USDA's) National Organic Program (NOP). Many countries outside the EU also legally protect organic products or are developing regulations for organic farming. These regulations all lay down minimum rules governing the production, processing and import of organic products, including inspection procedures, labelling and marketing. With respect to the World Trade Organization (WTO), the role of the Codex Alimentarius Guidelines (see Section 8.4.3) is increasingly related to the future harmonization of rules for organic production and questions of equivalence between countries.

8.4.1 EEC Regulation 2092/91 of the European Union

The EU started to develop a regulatory framework for organic farming in response mainly to pressure not only from consumer groups, but also from organic farming organizations. Many years of intensive discussion with the private sector, in particular IFOAM, took place before the regulation was published in 1991.

In EU member states, the labelling of plant products as organic is governed by EU Council (EEC) Reg. 2092/91, which came into force in 1992, while products from organically managed livestock are governed by EU Reg. 1804/99, which took effect in August 2000. They protect producers from unfair competition, and they also protect consumers from 'pseudo-organic' products. Plant and animal products and processed agricultural goods imported into the EU may be labelled as organic only if they conform to EU Regs 2092/91 and 1804/99. Each EU member state or associated European country is responsible for enforcement and for its own monitoring and inspection system. Applications, supervision and penalties are dealt with at the level of the member states of the EU. At the same time, each country is responsible for interpreting the regulations and implementing them in its national context. The EEC regulation has been amended many times, by updating or supplementing the technical annexes (Schmidt and Haccius, 1998). In March 2000, an EU Community logo for organic products was introduced.

Since 2004, there has been a debate regarding how Reg. 2092/91 should be revised and simplified by putting more emphasis on the basic principles of organic farming, as outlined in the European Action Plan (European Commission, 2004). In December 2005, a new draft proposal of an EU council regulation 'on organic production and labelling of organic products' for organic farming was published (European Commission, 2005a). After intensive discussion with the private sector, in June 2007 the EU council voted in favour of a new compromise text, which will be amended through implementation rules by the EU Commission in 2007. This new regulation should be in force by 2009.

8.4.2 US National Organic Program

Similar to the EU Reg. 2092/91, the USDA's NOP requires all products labelled as organic in the USA to meet the US standards, which took effect in October 2002 (see www.ams.usda.gov/nop/indexNet.htm). The US regulation is more specific than EU Reg. 2092/91 in its requirements for imports, and requires imported products to fully meet the NOP provisions. The US system accredits certification bodies as agents to operate the US certification programme published as part of the rule.

As of January 2006, as many as 93 certification bodies (53 US and 40 foreign) were accredited by USDA, and only products certified by them may be exported to the USA (Kilcher *et al.*, 2005).

8.4.3 Codex Alimentarius Guidelines

The need for clear and harmonized rules has been taken up not only by private bodies, IFOAM and state authorities (e.g. EU Reg. 2092/91), but also by the United Nation's Food and Agriculture Organization (FAO) and World Health Organization (WHO). In 1991, the Codex Alimentarius Commission, a joint FAO/WHO Food Standards Programme (with participation of observer organizations such as IFOAM and the EU) began developing guidelines for the production, processing, labelling and marketing of organically produced foods. In June 1999, the plant production guidelines and in July 2001, the guidelines for animal production were approved (Codex Alimentarius Commission, 1999/2001). In 2004, the criteria for new inputs as well as substances for processing were updated. These guidelines define the nature of organic food production and prevent claims that could mislead consumers about the quality of the product or the way it was produced. They take into account the current regulations in several countries, particularly EU Reg. 2092/91, as well as the private standards used by producer organizations, and especially the IFOAM Basic Standards (Schmid, 2002a).

The sections on plant and animal production are already well developed in the Codex Guidelines. Regarding processing, especially of animal products, there is an ongoing debate in the Codex Alimentarius Organic Working Group on limits on the use of food additives and processing aids. The revision work is taking into account consumer expectations for minimal processing and little use of additives, while allowing a range of products in different areas.

The Codex Guidelines are an important step in the harmonization of international rules to build consumer trust. They will be important for equivalence judgements under the rules of the WTO, and are important in giving guidance to governments in developing national regulations for organic food, particularly in developing countries and countries in transition.

These guidelines will be regularly reviewed at least every 4 years, according to a standard Codex procedure. Regarding the list of inputs, there is a possibility of an accelerated procedure that helps in updating the amendments more quickly. The new criteria for agricultural inputs, as well as those for additives and processing aids are used in such a way that decisions on future inputs are supported by technical submissions evaluated according to these criteria. Since 2004, there has been an ongoing discussion within Codex about the nature of the list of inputs. A majority of the Codex member states and observers from the private sector want to have an indicative, but very restrictive, list of substances that reflects a high worldwide consensus.

Further information about Codex Alimentarius is available at www.codexalimentarius.net. There is also a section on organic farming on the FAO web site: www.fao.org/organicag/. The Codex Alimentarius Guidelines on Organic Agriculture can be downloaded from ftp://ftp.fao.org/codex/standard/en/CXG_032e.pdf

8.5 Changes in the Role and Content of the Standards and Regulations

The early standards for organic farming typically were written more in the form of recommendations than standards, putting more emphasis on organic farming principles. Furthermore, many standards were formulated in a way that left room for regional and site-specific implementation. It is interesting that the nature and character of the standards have changed over time (Table 8.1).

One can say (at the risk of simplification) that in the pioneer phase the standards brought organic farmers together, whereas later the standards seemed to divide them. More and more private standard-setting organizations exist with differing standards, each claiming that it has additional or more detailed requirements beyond the common rules of the EU, USDA or other governmental regulations, but also beyond the private IFOAM Basic Standards (Schmid et al., 2007). These differing rules and the competition among labels might, on the one hand, stimulate the further development of organic agriculture, but could be confusing on the other, especially for occasional and less committed consumers of organic foods.

8.6 The Further Development of the Standards and Regulations in a Societal Context

The development of standards always reflected the general consciousness of the organic agriculture movement. This is seen in areas taken up and further developed in the IFOAM Basic Standards (Schmid, 1992; Padel et al., 2004). At first the main focus was on plant production. The basic idea was that a healthy soil would be the key to producing healthy crops and thus help to improve human health. In general, animal husbandry during this period was looked upon primarily as a means to improve plant production. In such a perspective, animals were mainly

Table 8.1. Content and function of standards for organic farming in different time periods.

Time period	Content of the private sector standards	Content of public regulations	Function of standards and regulations	Main actors
Early 1980s	General principles, conversion rules, mainly crop production rules (e.g. humus management, fertilization, plant protection), animal feeding, storage; first labelling rules	Generally none	Standards primarily to embody the more general principles in a more concrete form. Standards to give actors an identity (inside and outside the movement)	Mainly farmers and pioneer scientists; inspection mostly done by farmers
1985–1990	Elaboration of detailed rules for livestock (see Table 8.2) and processing. Preliminary criteria for the evaluation of certification bodies (IFOAM)	First governmental rules/logos (France). First conversion payments (e.g. Denmark)	Standards widened in scope, emphasizing the holistic approach of organic farming. More guarantees for the consumers with more inspection, certification and labelling rules	Farmers, advisors, farmers' associations
1991–1995	First debates about GMO (first a moratorium on their use, later prohibited). Development of a detailed set of criteria for the accreditation of certification bodies	Rewriting standards in legal language. More precise legal requirements for inspection	Standards acquire a triple function: – Guidance for farmers – Legal basis for inspection/certification – Basis for special payments to farmers Standards are the drivers of market development	Public administrators, e.g.: EU Commission (Reg. 2092/91), USDA. Start of Codex Alimentarius involvement. Start of IFOAM Accreditation Programme

Continued

Table 8.1. Continued.

Time period	Content of the private sector standards	Content of public regulations	Function of standards and regulations	Main actors
1996–2000	Development of new areas such as aquaculture, textiles, fibre production and plant breeding	Development of detailed animal husbandry rules (e.g. EU, Codex) More detailed rules for critical areas, e.g. use of copper	Rewriting standards for inspection bodies, making them more detailed and inspectable Standards for new areas helping to develop new markets	Stronger involvement of governments in setting the rules (e.g. EU, USDA) Private sector in new areas
2001–2005	New areas such as cosmetics Start of a discussion about basic principles New partnerships with fair trade and other sustainability label organizations	Mainly dealing with implementation rules to make regulations stricter (e.g. for feed, seed, processing)	Standards getting overly prescriptive – no longer really linked to the basic principles Inspection and certification rules seen as expensive and bureaucratic and as a strong barrier to conversion	Strong influence of governmental bodies New forms of cooperation among IFOAM, FAO and UNCTAD
2006–	Development of other instruments such as Code of Best Practice Further development of alternative certification systems (e.g. risk-based inspection, group certification)	Revision of state rules, making them more principle-based and accessible for developing countries	Need for standards to regain a function in promoting organic farming and not hindering it	New partnerships between private and public sectors

manure producers needed for the important process of composting. In addition, ruminants had the role of digesting forage legumes in order to permit a balanced crop rotation that built soil fertility and provided nitrogen for fertilization.

By the 1980s, animal production was receiving increasing attention under the influence of public debate, as summarized in Table 8.2. The protests against industrialized and polluting agriculture also included protests against 'factory farming', i.e. intensive livestock production based on purchased feeds. Ideas of more 'natural' and animal-friendly ways to raise livestock were embraced by organic farmers in developing alternatives to this (Padel *et al.*, 2004; see also Chapter 5, this volume, for a discussion of how public attitudes influenced the treatment of animals in organic farming).

Since 2000, livestock issues have remained the subject of public debate, e.g. foot-and-mouth disease, toxic residues in feeding materials and avian influenza. These no doubt will continue to be challenges for the organic agriculture movement.

As with the eventual development of animal standards, the IFOAM Basic Standards have been extended into other new areas. More recent extensions have been into aquaculture and textiles/fibres, in both cases first in 2000 as draft standards and then in 2005 as full standards.

8.7 Major Controversies and Issues of Debate

Over the years, many controversial issues have been debated with greater or lesser intensity within IFOAM and at the national level. Some important examples resulting in actual or potential changes include:

Conversion/transition time to full organic management: From the beginning there was a debate over how long a period of conversion should be required before the farm is considered organic. At first the issue of pesticide residues was dominant. Later, other criteria were used, such as the adaptation of the farm's agroecosystem, the farmer's understanding of organic farming (the learning process), the previous land management (such as whether it had been farmed traditionally with no use of chemical inputs), physiological processes in the case of animals, and administrative criteria for inspection (requiring at least one monitored period of a production cycle). These different viewpoints have led to different rules and interpretations in standards and regulations. For example, the IFOAM Basic Standards require at least a 12-month conversion period prior to the start of the production cycle, whereas with perennials (excluding pasture and meadows) a period of at least 18 months prior to harvest

Table 8.2. The historical development of organic livestock standards. (From Schmid, 2000.)

Time period	Public debate/Problem area	Views of critical consumers and producers	Influence on organic livestock standards
Before 1970	Deficiencies of micronutrients because of the intensive use of chemical fertilizers in food and feed	Healthy soil–healthy plants–healthy animals–healthy food. Animal manure an important source for fertilizing soil/plants	Animals should be an important part of an organic farm. There should be a balance between plant and animal production (biodynamic concept of 'farm organism')
1970	Problems with chemical residues in food (accumulation of organo-chlorides in the food chain, residues in mother's milk and animal products)	The risk of contamination by using conventional feed must be minimized	No prophylactic addition of antibiotics and hormones in feeds. Maximum allowances of feed from conventional sources (10–20% on dry matter basis)
1980	Problems of industrial animal production systems, animals suffering in intensive systems (battery chickens, etc.)	Conditions for 'happy' animals (animal welfare) have to be established, particularly on organic farms	Minimum requirements for outdoor access for all animals. Sufficient space in housing. No slatted floors; straw as bedding material; natural light
1994	Animals suffering during transport and slaughter	Animal welfare must include transport and slaughter	More detailed standards for transport and slaughter

Development of Standards

1994–1997	BSE crisis, debates about hormone use	Stricter regulations regarding feed ingredients from animals
1998	Problems with antibiotic resistance	Risks from feeding materials of animal origin must be eliminated
		The use of antibiotics must be much more strongly restricted (minimum of twice the official holding period, maximum of three courses of treatments with allopathic medicines).
	Risks for human beings from the therapeutic use of antibiotics to combat animal diseases or to promote growth must be reduced	
1998	GMO debate intensified	Prohibition of GM-derived components in the diet of organic livestock
	Risk of contamination of organic products through GMOs	
1999	BSE cases in continental Europe, resulting in second wave of the debate	Ruminants for meat production must be born on organic farms rather than undergoing a conversion period
	Risk of human infection through livestock	Further restrictions on feed ingredients

is required (IFOAM, 2005). EU Reg. 2092/91 calls for a conversion period of 3 years, which is broken down to a period of at least 2 years before sowing in the case of annual crops or 3 years before harvest for perennials.

Use of GMOs: In the IFOAM Basic Standards, genetically modified organisms (GMOs) and their derivatives were first excluded because their risks were not sufficiently known. Later (in late 1990s), their use was prohibited because they were considered incompatible with organic principles and were strongly rejected by organic food consumers.

Use of conventional seed: In 2000, stronger rules for organic seed were introduced by IFOAM because of the risk of GMO contamination. Organic seeds and plant materials of appropriate quality must be used when available. In 2004, the EU Commission required member states to provide databases for organic seeds.

Biodiversity: Since 2002, there has been a debate over whether more detailed standards should be elaborated for biodiversity and landscape amenities for organic farms, and if so, how.

Standards for social justice: Since 1990, when the first fair trade organizations were becoming more active, there was a debate over how far the organic agriculture movement should go in incorporating not only social standards but also fair trade principles (including fair prices) in its own standards (a question also discussed in Chapter 3, this volume). In the late 1990s, the IFOAM Basic Standards made reference to the Convention of the UN International Labour Organization relating to labour welfare and the UN Charter of Rights. However, the requirements remained rather vague. Only a few private standard-setting organizations have further developed these requirements, in partnership with other non-governmental organizations (NGOs).

Use of external inputs in plant production: There are different points of view regarding the use of inputs that are 'natural', but in some ways are more like synthetic inputs. An example is copper-based fungicide. In some regulations and standards the amount of copper that may be used for plant protection must be reduced to avoid soil contamination problems (e.g. in the IFOAM Basic Standards, since 2000). Furthermore, general national registration requirements for plant protection products often limit the use of some substances that are listed in organic farming standards but are not generally registered on the national level (Speiser and Schmid, 2004). For example, in several countries the costs for registering new plant protection products for organic farming are far too high compared with the potential market for such products.

Another debate concerning allowed external inputs, which goes back to 1979, was whether Chilean nitrate, a natural but highly soluble form of nitrate, is acceptable in organic farming (IFOAM, 1979). It was decided in 1982 that this natural source of nitrogen would be allowed only during conversion because it acts much like manufactured nitrogen fertilizer. Some years later, it was removed from the IFOAM Basic Standards both because it is similar to synthetic nitrogen fertilizer in its action and because it is not renewable.

Processing: Especially since 1990, there has been a strong focus on rules for organic processing because of the growing market for convenience products. There remains a controversy over which additives and processing methods can be allowed for different product groups without compromising the integrity of organic food. Some organic labels refer only to the raw materials, while for others it is important to have clear requirements for processing methods (Meier-Ploeger and Vogtmann, 1998; Schmid *et al.*, 2000, 2004).

8.8 Do the Standards Still Reflect the Basic Principles of Organic Farming?

The growing market, the emergence of large-scale organic production and the involvement of large conventional companies on a global scale are seen by some organic farmers and other organic farming activists as threatening organic farming's ability to function as an alternative to conventional agriculture (see Chapter 3, this volume, for a discussion of this question). In particular, they question whether the principles underlying the first IFOAM Basic Standards (1980) are still followed:

1. To work as much as possible within a closed system, and draw upon local resources;
2. To maintain the long-term fertility of soils;
3. To avoid all forms of pollution that may result from agricultural techniques;
4. To produce foodstuffs of high nutritional quality and sufficient quantity;
5. To reduce the use of fossil energy in agricultural practice to a minimum;
6. To give livestock conditions of life that conform to their physiological needs and to humanitarian principles;
7. To make it possible for agricultural producers to earn a living through their work and develop their potentialities as human beings.

Since 1980, these principles have been changed several times. They were reformulated not necessarily to make them more enforceable, but rather as a way of adapting them to the realities of the globalized market.

In particular, the first principle, to work as much as possible within a closed system and draw upon local resources, was changed in a way that some early activists regarded as watering it down. In 2002, instead of 7 principles, the IFOAM Basic Standards put forward 15 principal aims (Box 8.1).

There is therefore a growing and renewed interest in values and principles of organic farming that could guide future development.

Box 8.1. The principal aims of organic production and processing. (From IFOAM, 2002.)

Organic production and processing is based on a number of principles and ideas. All are important and this list does not seek to establish any priority of importance. The principles include:
- To produce sufficient quantities of high quality food, fibre and other products;
- To work compatibly with natural cycles and living systems through the soil, plants and animals in the entire production system;
- To recognize the wider social and ecological impact of, and within, the organic production and processing system;
- To maintain and increase long-term fertility and biological activity of soils using locally adapted cultural, biological and mechanical methods as opposed to reliance on inputs;
- To maintain and encourage agricultural and natural biodiversity on the farm and surrounds through the use of sustainable production systems and the protection of plant and wildlife habitats;
- To maintain and conserve genetic diversity through attention to on-farm management of genetic resources;
- To promote the responsible use and conservation of water and all life therein;
- To use, as far as possible, renewable resources in production and processing systems and avoid pollution and waste;
- To foster local and regional production and distribution;
- To create a harmonious balance between crop production and animal husbandry;
- To provide living conditions that allow animals to express the basic aspects of their innate behaviour;
- To utilize biodegradable, recyclable and recycled packaging materials;
- To provide everyone involved in organic farming and processing with a quality of life that satisfies their basic needs, within a safe, secure and healthy working environment;
- To support the establishment of an entire production, processing and distribution chain which is both socially just and ecologically responsible;
- To recognize the importance of, and protect and learn from, indigenous knowledge and traditional farming systems.

A broader debate was initiated in 2004 by an IFOAM task force on the one hand, and in a parallel process by a new EU project on the revision of EU Reg. 2092/91 (see www.organic-revision.org) on the other. During these discussions it became clear that the existing definitions of organic farming and the principal aims as formulated in the IFOAM Basic Standards and other standards are no longer necessarily adequate as a guide for sustainable and dynamic development (details about the discussion process are available at: www.ifoam.org).

Formulating strong principles is seen by many as a tool to evaluate the development, correct the course and avoid unwanted consequences by way of timely care. In particular, there is a belief that basic ethical principles can support the interpretation and development of organic standards in existing areas of conflict and in new areas of production, processing and traceability. Moreover, they can serve directly as a guide for organic practices in areas where standards are hard to set. Thus, by relieving the rules of some of their functions, it is hoped that the present trend towards evermore complicated rules can be stopped and reversed (Alrøe *et al.*, 2005).

The broad discussions of the development of overarching basic principles that both IFOAM and the EU organic revision projects held in 2004 and 2005 resulted in the final acceptance of four main principles at the IFOAM General Assembly in Adelaide, Australia, in 2005 (Box 8.2).

Although these principles have strong support worldwide, several concerns remain that should be taken into account. For some people, these principles are still too general and can only be used for guidance, not for concrete decisions. Therefore, they might just be a starting point, eventually transformed into working principles and aims. If this is done, the principles can be translated into decision criteria, for example, for the evaluation and acceptance of new inputs for crop production, as

---

Box 8.2. The four IFOAM main principles of organic production. (From IFOAM, 2005.)

The Principle of Health – Organic agriculture should sustain and enhance the health of soil, plant, animal and human as one and indivisible.
The Principle of Ecology – Organic agriculture should be based on living ecological systems and cycles, work with them, emulate them and help sustain them.
The Principle of Fairness – Organic agriculture should build on relationships that ensure fairness with regard to the common environment and life opportunities.
The Principle of Care – Organic agriculture should be managed in a precautionary and responsible manner to protect the health and well-being of current and future generations and the environment.

was elaborated within an EU project (see www.organicinputs.org). Such working principles should take into account the differences in values among different countries and different groups of actors. Procedures should be developed for balancing such differences through an open, transparent and participatory process.

Whereas, the current EU Reg. 2092/91 has no real overarching principles (although it will have them in the new regulation to be put in place in 2009), the Codex Alimentarius Guidelines (CAC/GL 32/2001) developed the following set of main principles in 2001, with strong involvement of the private sector (IFOAM).

An organic production system has several design goals:

- To enhance biological diversity within the whole system;
- To increase soil biological activity;
- To maintain long-term soil fertility;
- To recycle wastes of plant and animal origin in order to return nutrients to the land, thus minimizing the use of non-renewable resources;
- To rely on renewable resources in locally organized agricultural systems;
- To promote the healthy use of soil, water and air as well as minimize all forms of pollution thereto that may result from agricultural practices;
- To handle agricultural products with emphasis on careful processing methods in order to maintain the organic integrity and vital qualities of the product at all stages;
- To become established on any existing farm through a period of conversion, the appropriate length of which is determined by site-specific factors such as the history of the land, and type of crops and livestock to be produced.

It will be interesting to see how these different principles in different standards and regulations will be used in the implementation process in the future.

8.9 Development of Inspection, Certification and International Accreditation

The organic market is confronted with dozens of private sector standards and governmental regulations. Although many of the differences among these standards are minor, there is a need for harmonization. Mutual recognition and equivalency among these systems are very limited. Particularly for producers in developing countries, the existence of numerous systems of inspection, certification and accreditation are a major obstacle to the continuing and rapid development of the organic agriculture sector. Certification has become a big business, with some key players active worldwide.

8.9.1 Certification and assessment of conformance with regulations

In the beginning of standards development, private certification bodies confirmed to consumers through their certification logos that the farm or production unit and its products fulfilled the organic production standards as a guarantee for quality. In several countries there is still a high consumer confidence in these private labels of the certification bodies and labelling organizations. These certification labels are generally registered as trademarks. In other countries, private certification bodies and private logos lost importance when a state system and logo were introduced, as in Denmark or France.

With EU Reg. 2092/91, which came into force in 1992, a European-wide state-based system for determining conformity with the regulation was introduced, and other countries have generally followed this approach in establishing their regulations. A key element of this system is that it allows the recognition of private certification bodies by a designated 'competent authority' according to specified criteria. There are some small differences in how countries handled this. In many cases a certification body must comply with the criteria of the EN 45011 Norm of the EU or ISO/IEC Guide 65 (Guide of the International Standardization Organization for Certification Bodies), as is required by the EU, for example (European Commission, 2005b). However, not all countries make reference to the ISO 65 Guide; some have chosen to base their requirement on IFOAM criteria (Michaud *et al.*, 2004).

8.9.2 Private accreditation

Starting in 1983 there was discussion within IFOAM about developing a system of recognition among certification bodies on a private basis. The first initiative for an international system was proposed by Look uit het Brook of the Netherlands (who at that time was chair of the IFOAM Technical Committee and also an organic farm adviser), together with market actors from Belgium, such as Carl Haest. This proposal for a private international certification system in close collaboration with IFOAM under the name 'Qualitree' was, however, not supported by the IFOAM Board.

In 1985, the (former) IFOAM Technical Committee started a process of evaluation of certification bodies who were IFOAM members to gain experience regarding recognition among certifiers. Already in 1986 several reports based on self-elaborated criteria were made by the Technical Committee. Later IFOAM developed the Accreditation Programme, which was approved in 1990 and began work in 1992. In 1993, detailed criteria for organic certifiers were published for the first time, and in 1995 the first certification bodies were accredited. In 1997, the International Organic Accreditation Service (IOAS) was formally established as an independent body, and in 1999 the IFOAM Accredited Certifier seal was launched. In 2000, a multilateral agreement was made among the accredited certifiers to accept each other's certification on the basis of the common IFOAM Basic Standards and criteria for accreditation. By 2005, a total of 33 certification bodies was accredited (Bowen, 2005).

8.9.3 Cooperation between the private and public sectors

With the development of the international market, the issue of harmonization is becoming increasingly important (Hamm et al., 2002). Among the most significant future framework considerations in the context of harmonization will be the WTO agreement on the application of Sanitary and Phytosanitary Measures and Technical Barriers to Trade, where reference is made to international standards (such as the Codex Alimentarius Guidelines, but also the IFOAM Basic Standards). This agreement will be particularly important in the case of trade disputes.

For many inspection/certification bodies it does not make sense to participate in a private as well as a public system, as this implies high costs for accreditation. In 2001, IFOAM, FAO and UNCTAD decided to join forces in finding solutions. Several conferences took place and an International Task Force on Harmonization and Equivalence in Organic Agriculture was launched that has continued working. The Codex Alimentarius Guidelines and the IFOAM Basic Standards are a basis for such a process (Michaud et al., 2004).

8.10 Challenges for the Future

Several future challenges are related to standard setting:

- Many certifiers feel that they are overburdened and have lost control as more and more standards, regulations and inspection requirements are imposed. They do not want more standards and regulations, but would prefer to rely more on the basic principles of organic agriculture. The ongoing debate about these basic principles

that IFOAM initiated in 2004 is highly relevant to the further development of organic farming, as it can provide guidance and decision criteria for the further development and even for a simplification of standards.
- Within standards and inspection/certification procedures a stronger focus on risk factors is needed. For example, it might be appropriate to make more inspections per year in partially converted operations with many different products (and therefore a higher potential risk of fraud or mixing) than in small traditional farms with only one product group.
- Differences in standards create problems only when they create competitive disadvantages, lead to consumer distrust or affect the integrity of the organic farming system (i.e. they conflict with basic principles of organic farming).
- To achieve more regionally adapted organic farming it is important to allow regional variation in the standards. This needs clear principle-based criteria for different areas such as input evaluation, animal-friendly husbandry and careful processing with minimal use of additives.
- The aim should be that standards are equivalent, not necessarily identical. Farmers as well as others in the organic food sector should have a clear but not overly prescribed regulatory framework that allows them to be innovative and creative and gives them a feeling of self-responsibility.
- There is a need to reflect more about public–private partnerships in the harmonization and implementation of the rules of organic farming on a worldwide as well as on a national level. The ongoing process of cooperation among IFOAM, Codex Alimentarius and UNCTAD in the harmonization of standards is important, but should always focus on the overall aims of organic farming, and not lead to even more bureaucracy.
- Harmonizing the multi-accreditation and certification requirements remains a major challenge for organic agriculture, yet the growing costs and regulatory burdens placed on small farmers must also be expeditiously and equitably resolved. The fact that neither the USA nor the EU has formally recognized the leading non-governmental third-party accreditation system, the IOAS, reflects a major hurdle that must be overcome if 'market rationalization' is to take place. The goal is for the system to be driven by and transparent to as broad a base of stakeholders as possible (Sligh and Christman, 2003).
- There is also a need to focus more on the expectations that consumers have with regard to the wider issues of organic agriculture, such as fair trade, social justice, biodiversity benefits, landscape amenities, environmental management along the whole food chain

and animal welfare. If consumers' expectations diverge too much from reality, it can create major problems for the image of organic farming. Therefore the standards in some areas should be more clear and precise about what must be achieved. This does not necessarily mean more standards; other areas might be simplified.
- Food safety issues such as pesticide residues, GMO contamination and microbiological risks will remain important for many consumers. Standard-setting organizations will have to take these concerns into account without losing the process-based approach of organic farming, e.g. with better monitoring of food safety issues and well-harmonized procedures in case problems do occur (Schmid, 2002b).

References

Alrøe, H., Schmid, O. and Padel, S. (2005) Ethical principles and the revision of organic rules. *The Organic Standard,* June 2005.

Aubert, C. (1970) *L'Agriculture biologique.* Le courrier du livre, Paris.

Bowen, D. (2005) Presentation at General Assembly of IFOAM on Organic Guarantee Systems. Adelaide, Australia.

Codex Alimentarius Commission (1999/2001) *Guidelines for the Production, Processing, Labelling and Marketing of Organically Produced Foods.* CAC/GL 32-1999/Rev 1 – 2001, Rome. Available at: http://www.codexalimentarius.net/download/standards/360/CXG_032e.pdf

De Silguy, C. (1991) *L'Agriculture biologique.* Presses universitaires de France, Paris.

European Commission (2004) European Action Plan for Organic Food and Farming. Brussels 10 June 2004. COM(2004)415 final. Available at: europa.eu.int/comm/agriculture/qual/organic/plan/comm_en.pdf 07.07.04

European Commission (2005a) Proposal for a Council Regulation on organic production and labelling of organic products (21 December 2005). Available at: http://ec.europa.eu/agriculture/qual/organic/index_en.htm

European Commission (2005b) Consolidated version of the Council Regulation (EEC) No. 2092/91 of June 1991 on organic production of agriculture products and indications referring thereto on agricultural products and foodstuffs. CONSLEG: 1991R2092. Available at: europa.eu.int/eur-lex/lex/LexUriServ/site/en/consleg/1991/R/01991R2092-20051005-en.pdf

Fischer, R. (1982) Der andere Landbau. Dissertation ETH Nr. 6636: Das Selbstbild von biologisch wirtschaftenden Bauern. Verlag Madliger Schwab, Zürich.

Hamm, U., Gronefeld, F. and Halpin, D. (2002) *Analysis of the European Market for Organic Food.* Organic Marketing Initiatives and Rural Development Series, Vol. 1. School of Business and Management, University of Wales, Aberystwyth, UK.

Heinze, H. (1972) Internationale Demeter-Tagung hilft gleiche Qualitätsrichtlinien sichern. *Lebendige Erde* Jg. 1972. Nr. 4, p. 148.

IFOAM (1978) Summary of the minutes of the Bi-Annual General Assembly held on 6–8 October 1978. Internal IFOAM document, Montreal, Canada.

IFOAM (1979) The basic rules of biological agriculture standards under consideration by IFOAM. Internal discussion paper for IFOAM Member Organizations. Secretariat of IFOAM Technical Committee Oberwil/BL/(Switzerland).

IFOAM (1980) Recommendations for international standards of biological agriculture. General Assembly. IFOAM Secretariat, Topsfield, Massachusetts.

IFOAM (1982) Standards of biological agriculture for international trade and national standards – with restricted validity to 2 years. IFOAM Secretariat, Topsfield, Massachusetts.

IFOAM (2002) *IFOAM Norms – IFOAM Basic Standards – IFOAM Accreditation Criteria.* IFOAM, Tholey-Theley, Germany.

IFOAM (2005) The IFOAM Norms for organic production and processing including IFOAM Basic Standards and IFOAM Accreditation Criteria. Version 2005. IFOAM, Bonn, Germany.

Jespersen, L.M. (1998) International and national 'organic' standards in the EU. Report and database in Excel. Section for Ecology, Danish Agricultural Advisory Centre, Denmark.

Kilcher, L., Huber, B. and Schmid, O. (2005) Standards and regulations. In: Willer, H. and Yussefi, M. (eds) *The World of Organic Agriculture.* FiBL, Frick, Switzerland, pp. 74–83.

Koepf, H.H. and von Plato, B. (2001) *Die biologisch-dynamische Wirtschaftsweise im 20. Jahrhundert.* Verlag am Goetheanum, Dornach, Switzerland.

Meier-Ploeger, A. and Vogtmann, H. (1998) *Dokumentation der Verarbeitungsrichtlinien für Produkte aus ökologischem Anbau; Teil 1: Arbeitspapier und Aktionsplan.* Kassel University, Witzenhausen, Germany.

Michaud, J., Wynen, E. and Bowen, D. (2004) *Harmonization and Equivalence in Organic Agriculture,* Vol. 1. International Task Force on Harmonization and Equivalence in Organic Agriculture. FAO/IFOAM/UNCTAD, Rome/Bonn/Geneva.

Nature et Progrès (1974) Commercialisation des produits biologiques. *Cahier Techniques* 2/1974. E 03. St. Geneviève-des-Bois, France.

Padel, S., Schmid, O. and Lund, V. (2004) Organic livestock standards. In: Vaarst, M., Roderick, S., Lund, V. and Lockeretz, W. (eds) *Animal Health and Welfare in Organic Agriculture.* CAB International, Wallingford, UK, pp. 57–72.

Scheidegger, W. (1993) *Biologischer Landbau – Illusion oder Chance?* Zentrum für biologischen Landbau, Grosshöchstetten, Germany.

Schmid, O. (1992) IFOAM's Basic Standards – 15 years work of the Standards Committee of IFOAM. In: *20 Years of IFOAM.* International Federation of Organic Agriculture Movements, Tholey-Theley, Germany, pp. 20–21.

Schmid, O. (2000) Comparison of European Organic Livestock Standards with national and international standards – problems of common standards development and future areas of interest. In: Hovi, M. and Trujillo, R.G. (eds) *Diversity of Livestock Systems and Definition of Animal Welfare. Proceedings of the Second NAHWOA Workshop, Córdoba, 8–11 January 2000.* University of Reading, Reading, UK, pp. 63–75.

Schmid, O. (2002a) Comparison of EU Regulation 2092/91, Codex Alimentarius Guidelines for Organically Produced Food 1999/2001, and IFOAM Basic Standards 2000. In: Rundgren, G. and Lockeretz, W. (eds) *Reader, IFOAM Conference on Organic Guarantee Systems, International Harmonisation and Equivalence in Organic*

Agriculture, 17–19 February 2002, Nuremberg, Germany. IFOAM, Tholey-Theley, Germany, pp. 12–18.

Schmid, O. (2002b) The food safety debate and development of standards/regulations for organic farming and organic food. In: *Proceedings of the 14th IFOAM Organic World Congress: Cultivating Communities, 21–24 August 2002, Victoria, Canada*. Canadian Organic Growers, Ottawa, p. 213.

Schmid, O., Blank, C., Halpin, D. and Bickel, R. (2000) Evaluation of governmental regulations and private standards for processing of organic food products – overview on some national standards, on IFOAM Basic Standards, Codex Alimentarius Guidelines and EU-Regulation 2092/91. In: Stucki, E. and Meier, U. (eds) *Proceedings 1st International Seminar 'Organic Food Processing', 28–29 August 2000, Basel*. Hochschulverlag AG an der ETH Zürich, Switzerland, pp. 59–67.

Schmid, O., Beck, A. and Kretzschmar, U. (2004) *Underlying Principles in Organic and 'Low-Input Food' Processing – Literature Survey*. First publication within the EU IP Project 'Quality Low Input Food'. FiBL, Frick, Switzerland. 112. Available at: www.qlif.org

Schmid, O., Huber, B., Ziegler, K., Hansen, J.G., Plakolm, G., Gilbert, J., Jespersen, L.M., Lomann, S., Micheloni, C. and Padel, S. (2007) *Analysis of EEC Regulation 2092/91 in Relation to Other National and International Organic Standards*. Final Deliverable D3.2 EEC 2092/91 (Organic) Revision. Frick, FIBL. Research Institute of Organic Agriculture. Organic Revision Project website: www.organic-revision.org

Schmidt, H.P. and Haccius, M. (1998) *EU Regulation 'Organic Farming'. A Legal and Agro-Ecological Commentary on the EU's Council Regulation No. 2092/91*. Margraf Verlag, Weikersheim, Germany.

Sligh, M. and Christman, C. (2003) *Who Owns Organic? The Global Status, Prospects, and Challenges of a Changing Organic Market*. Rural Advancement Foundation International – USA, Pittsboro, North Carolina. Available at: www.rafiusa.org/pubs/OrganicReport.pdf

Soil Association (1967) First Soil Association standards for organically produced food. *Mother Earth*. Available at: http://www.soilassociation.org/web/sa/saweb.nsf (Archive)

Speiser, B. and Schmid, O. (eds) (2004) *Current Evaluation Procedures for Plant Protection Products Used in Organic Agriculture*. Proceedings of a workshop held 25–26 September 2003 in Frick, Switzerland as part of the EU Concerted Action Project Organic Inputs Evaluation. Available at: www.organicinputs.org

Vogt, G. (1999) Entstehung und Entwicklung des ökologischen Landbaus im deutschsprachigen Raum. PhD dissertation, Justus-Liebig-Universität, Giessen, Germany.

IFOAM and the History of the International Organic Movement   9

B. Geier

Director, Colabora, Alefeld 21, 53804 Much, Germany*

9.1 Introduction

A worldwide network known as the International Federation of Organic Agriculture Movements (IFOAM) today unites about 750 member organizations and institutions in over 100 countries around the globe. Its main task is to represent and coordinate the full diversity of the organic movement. Besides being united at the global level and across all sectors, IFOAM's members are also organized in regional groups, e.g. European Union, Mediterranean and Asia, as well as specific interest groups, e.g. the Forum of Consultants, the Aquaculture Group and the Organic Trade Forum. The multifold activities of IFOAM are also carried out through committees, working groups and task forces. All these components are supported by a professional staff in the Head Office in Bonn, as well as people working in Uganda, the USA and Italy.

Among the aims and activities of IFOAM are:

- Information exchange about all facets of organic agriculture;
- Promotion of the worldwide development of organic agriculture;
- Providing common platforms for interest groups;
- Exchange of knowledge and experience among members and the organic movement as a whole;

---

* When this chapter was written, Bernward Geier was Director of International Relations and Marketing, International Federation of Organic Agriculture Movements.

©CAB International 2007. *Organic Farming: an International History* (Lockeretz)

- Representing the organic movement in international institutions and agencies;
- Continuously revising the IFOAM Basic Standards and the Norms for Accreditation (IFOAM, 2002);
- Developing a harmonized international organic guarantee system from the basic standards to the IFOAM Accreditation Programme.

With the organic movement united in IFOAM, driven by dynamic activists and built on solidarity, organic agriculture has a common voice, ensuring that its full potential to solve agricultural problems gets worldwide attention and recognition.

9.2 The Roots of the Organic Movement

There once was a philosophy of agriculture that said that it should be as much in harmony with nature as possible. This idea is deeply rooted in ancient agriculture as it was and to a considerable extent is still practised in places such as India, China and the Andes. Organic agriculture reflects this general philosophy, but the recent history of concepts such as organic, biodynamic, agroecological and natural farming, and other related concepts, can more specifically be traced back to early in the 20th century. It is appropriate to remember that the organic and biodynamic philosophies were quite far reaching, and not simply 'anti-chemical', i.e. opposed to the use of synthetic pesticides and fertilizers. (For example, when Rudolph Steiner gave his course in 1924 that established the foundations of biodynamic agriculture, chemical inputs were largely unknown.) It was the organic pioneers' concern for soil fertility and their early realization of the problems arising from agriculture's shift towards monoculture and industrialization that inspired them to develop alternatives (see Chapter 2, this volume).

The history of the organic movement has a clear and logical sequence. First came the philosophy and teachings, which were based on observation of nature and respect for natural laws. In turn, the organic pioneers transformed these principles into practical farming methods. Finally there emerged a worldwide organic movement.

More accurately, it was not one homogeneous movement, but different schools of thought that created a diversity of national and regional movements and organizations. Eventually these movements and organizations felt the need to coordinate their activities on an international scale. This need was addressed by founding IFOAM. This effort has a precise birthdate – 5 November 1972 – and a very historic birthplace, Versailles. The five founding organizations were the Soil Association (UK), the Swedish Biodynamic Association, the Soil Association of

South Africa, Rodale Press (USA) and Nature et Progrès (France), the organic farmer organization that initiated it.

Yet it was not so much these organizations that stood at IFOAM's cradle as it was individuals. First to be mentioned is Roland Chevriot of France, who was the main initiator of IFOAM's founding, encouraged by Bob Rodale. Also at the founding assembly were Jerome Goldstein, representing the Rodale Press, and Kjell Armani, representing the Swedish Biodynamic Association. But it was actually four women who not only stood at the cradle but also started to rock the baby. Foremost among them was a true pioneer, and probably the most influential: Lady Eve Balfour (representing the Soil Association). Also involved were Pauline Raphaely from the Soil Association of South Africa, Mary Langman from the UK, and Karin Mundt, who was travelling the world as journalist for a French magazine and thus was best positioned to start the networking.

The time of IFOAM's founding was not long after the famous year '1968'. At the time organic farming was anti-establishment, if not absolutely revolutionary. This spirit was reflected in the early days of the federation, when no minutes or records were kept and no hierarchical structure or positions were wanted or established.

A young Frenchman, Denis Bourgeois, was entrusted not only to help organize the founding assembly, but also to nurse the infant organization. His memories of how all this came about convey the feeling of the IFOAM's founding more than 30 years ago. In IFOAM's 25th anniversary publication, he described how it all started:

> My first contact with the idea of IFOAM was in June 1972. I had just completed a BA in a business university but was mainly interested in alternative economies and the growing concern for ecology. Indeed, my dissertation had been on organic agriculture in France. Therefore I did not get, nor look for, a position offered from Procter & Gamble, Renault and the like, as most of my colleagues did. Instead I took up one from Roland Chevriot. Roland had been appointed president of Nature et Progrès, some two years before.
>
> It was not a conventional recruitment interview. The headquarters of the association were located in his modest house, surrounded by a 1 ha piece of land in a Parisian suburb. My first view of him that day was of two bare feet and a bit of leg. The rest of his body was hidden under an old Citroen 2CV that he was repairing. Roland was an engineer in Paris during weekdays, and a gardener and an organic activist during the evenings and at weekends. When I could see more of him, we went into a small room full of papers and books which was then Nature et Progrès' main office. He proposed I help organize a big national organic conference which was to take place in November, and also to work on the launching of an international federation – the conference providing the opportunity for a first meeting of representatives from the other organizations. He could not pay me for that job but he was going to rent the neighbouring house, so as to provide accommodation for the few people who like me had the job of

running the conference and his association. The association would also take care of our food, at least that which we could not grow ourselves in the surrounding garden. I said OK – and thus the first IFOAM volunteer was hired.

(Bourgeois, 1997, p. 12)

9.3 The Early Development of IFOAM and its Worldwide Network

Radical (in its original sense of going to the roots), innovative and filled with enthusiastic dedication, IFOAM's cornerstone was laid in 1972. Building upon the five founding member organizations, by 1975 it had already grown to 50 member organizations from 17 countries. Development then continued comparatively slowly, after 15 years reaching about 100 member organizations from 25 countries.

In 1976, the budget was only US$6,000; clearly people, not money, built the federation. Budget constraints remained a central problem, becoming very serious in the early 1980s, when only about US$11,000 was available.

The growth of IFOAM and the international network gained enormous momentum around 1986/87, when a loan from its then outgoing General Secretary Gunnar Videgard allowed IFOAM to employ its first full-time staff member. With a permanent employee and a General Secretariat combined with a very engaged World Board, the foundation was now laid for very dynamic and rapid development. Within just a few years the membership grew to 500 organizations in 70 countries. This healthy growth continues, with the most dynamic growth in recent years occurring in the developing world.

Much has changed since 1972. At the time, 'organic' people were not only radical – they also were seen by many as eccentric or even crazy. Part of the success of the organic agriculture movement lies in the fact that these negative attitudes have changed dramatically.

Yet looking at the early history of IFOAM, the spirit of the late 1960s is unmistakably reflected in how it started to develop its structure. A hierarchical structure clearly was not favoured. Instead, it was decided that the federation would have no president, but rather people nominated or elected by each group or committee to be responsible for each mission or action. But to be registered under French law as a non-profit organization, a president or at least three responsible persons would need to be named. Obviously, IFOAM chose the second option, naming Claude Aubert as General Secretary, Roland Chevriot as Treasurer and Denis Bourgeois as Administrator of the General Secretariat (all from France). IFOAM has been democratically structured ever since, with major decisions taken at the General Assembly, which is also where its governing body, the World Board, is elected.

9.4 Organizational Change

For over 30 years IFOAM has been travelling a long and winding road. Yet this road has always led upward and forward. Along the road were significant challenges, which were always taken to be opportunities. Some interesting and important advances along this road are worth reflecting upon.

In the early days, the movement was very much driven by scientists. The early scientific conferences were linked with the general assemblies and became a kind of continuing platform for the further development of IFOAM. Yet from the beginning the practical farming side and farmer organizations were also well represented, with three of the five founding members being farmer groups. Therefore, even in its initial phase IFOAM already was marching on two legs: organic farming and science. Many more 'legs' joined the organization over the years: people and companies active in publication, traders, processors, certification organizations, training institutions, national networks and many others. With the expansion to member organizations in developing countries, groups concerned with rural development and women also found their place in IFOAM.

A challenge that for quite some time was seen by many pioneers (the more dogmatic ones?) not as an opportunity but more as a threat was the trade sector's interest in becoming part of the worldwide organic movement. In the 1980s and early 1990s there were intense debates that showed a fear that the movement, with all its ideals and principles, could be 'taken over' by the economic power of business and industry. It took quite an educational process to get the grassroots membership to allow organic businesses to join IFOAM with the same services and privileges. In 1992, the World Board offered a proposal to open IFOAM to the industry (actually some organic companies were members from the early days). After heavy debates until 2 a.m., the proposal was defeated in the General Assembly. Two years later, a compromise was struck that distinguished not between non-profit (non-governmental organizations, NGOs) and for-profit (business) members, but rather between those institutions whose main activities, including making money, respectively are or are not in the organic sector. Thus, the only distinction between IFOAM members and associates since then has been that only those who are predominantly engaged in organic activities have voting rights. This compromise was an important milestone in the development of IFOAM because it opened the organization to the sector that profits the most from organic farming and thus can also contribute significantly to the international movement (e.g. via the membership fees).

Probably the most important milestone was the establishment of a permanent office and a hired staff. After more than a decade of hoping, wishing and planning, a first step was made in 1986, when a full-time, professional General Secretariat was set up. Starting with one staff position, today a total of about 25 people work for IFOAM (including the staff of the independent but affiliated International Organic Accreditation Service). This capacity building and development was possible only because the load continued to be carried substantially by volunteer workers and dedicated activists who have given and continue to give so much to IFOAM and the movement.

Looking at these milestones, there were only a few big steps in the development of IFOAM. It was the many daily advances and small jumps that had for over 30 years allowed IFOAM to maintain its leadership role and be the unifying institution for the organic movement all over the world.

9.5 IFOAM Reaches Out

From its beginning IFOAM realized that its main aim, promoting networking and exchange of knowledge, is most creatively achieved by bringing people together. Soon the biennial general assemblies were combined with a series of international scientific conferences, starting in 1977 in Sissach, Switzerland. Remarkably, with the first conference title 'Towards a Sustainable Agriculture', the organic movement introduced a concept that was already put into practice on many farms, but which only years later gained attention in the mainstream. The international scientific conferences continued to be held every 2 years, serving as the focal point and social highlight of the organic movement. Conference sites were as far-flung as Montreal, Witzenhausen (Germany), Ouagadougou (Burkina Faso), São Paulo, Christchurch (New Zealand) and Budapest.

Starting in 1985, IFOAM also offered more specific meeting opportunities by establishing a series of conferences on topics such as non-chemical weed control, trade, tea, coffee, cotton and biodiversity. The current calendar of events impressively proves that bringing people together is a main activity and strength of IFOAM. For example, in 2004 it organized three international conferences of its own: on seeds, coffee and biodiversity, respectively, and was a partner or participant in a total of 20 conferences, fairs and other events.

New platforms to bring people together have been developed through IFOAM's involvement in organic trade fairs. Its cooperation and patronage of the world's leading organic fair, BioFach, which takes place every February in Nuremberg, and its 'offspring' fairs in Rio de Janeiro, Washington, DC, and Tokyo, especially strengthened IFOAM's leadership role and provided new and exciting opportunities to foster the development of organic market.

From its beginning, IFOAM was also an active publisher. From a 'handmade' newsletter, an English- and a German-language magazine developed that were originally known as IFOAM Bulletins, today enjoying a readership in some 110 countries as the magazine *Ecology and Farming*. Central for exchanging information on internal affairs is the IFOAM newsletter. Actually to call it a 'newsletter' is misleading, considering that special issues sometimes run to 70 pages or more. The publication for members now has a name that reflects well the daily life of the federation: *IFOAM – In Action*. IFOAM also has moved with the times regarding modern communication technologies. For several years its web site (available at: www.ifoam.org) has been an important promotional tool and information source. Internal communication includes an intranet platform, and important action information goes out monthly to the members via the electronic newsletter *Insider*.

With the publication of IFOAM's first conference proceedings (Besson and Vogtmann, 1978), the tradition started that all IFOAM conferences would be documented in publications, thus making information available far beyond the people able to participate in the events. A big success, not only in its worldwide circulation but also as a source of income, has been the IFOAM directory *Organic Agriculture Worldwide*. Launched in the late 1980s, it now serves as the international 'yellow pages' of the organic movement.

In recent years, the range of IFOAM publications has expanded far beyond conference proceedings. Today about 50 publications include not only books and other printed material, but also CD-ROMs, DVDs and videotapes. English remains the predominant language, but IFOAM publications are also increasingly available in French and especially Spanish. Popular publications like the information brochure on genetic engineering have been translated into about a dozen languages, including Albanian, Portuguese and Sanskrit.

9.6 Building and Harmonizing the Organic Guarantee System

With international trade in organic products, which dates back to the 1970s, came a felt need to define on the international level what organic agriculture is all about, and to develop an organic guarantee system, including inspection, certification and ultimately accreditation.

In the eventual development of the organic guarantee system, first came a philosophy and concept of organic agriculture. Building on this, organic practices were developed on farms around the world. Only afterwards did organizations start to define organic agriculture through what nowadays are called standards or guidelines. IFOAM's engagement and leadership role in the organic guarantee system started in the late 1970s with the establishment of a committee that drafted the first IFOAM Basic Standards. Ever since, the standards have continually been discussed, expanded and updated by IFOAM's members. The important influence that IFOAM's work had and still has through its basic standards is seen in the many organic movements in numerous countries whose standards have been developed starting from the foundation that the IFOAM standards provide. Their importance is also shown by their having been translated into about 20 languages, including Chinese, Russian, Japanese and even Swahili.

Even more important has been the influence of the basic standards on governmental regulations for organic agriculture, starting with the development of the EU regulation of 1991 (see Chapter 8, this volume). All subsequent organic laws and regulations in the world have directly or indirectly built on the expertise and foundations provided by the IFOAM Basic Standards. This includes their influence on the international Codex Alimentarius Guidelines developed by the Food and Agriculture Organization (FAO) and the World Health Organization (WHO) of the United Nations.

With the IFOAM Basic Standards as a platform, all over the world a diversity of certification programmes has created an organic guarantee system that gives consumers confidence that they are buying genuine organic products. But such standards left a gap that IFOAM's membership eventually asked it to fill. Although there were common standards and professionally managed certification programmes, nobody verified the quality and performance of those programmes. This led to the development of the IFOAM Accreditation Programme, starting with a motion at the IFOAM General Assembly in Santa Cruz (USA) in 1986. The programme was officially launched in 1992; administered by the independent International Organic Accreditation Service, it is a central element in the IFOAM Organic Guarantee System (www.ioas.org). Today, 32 certification programmes and organizations active in over 70 countries are IFOAM-accredited. These certification programmes cover an estimated two-thirds of international organic trade.

Developing this system must be considered a success story. However, the diversity of certification programmes (and competition among them) and the engagement of governments (some consider it a takeover) in regulating the organic sector also led to confusion and trade obstacles, especially for smallholders and cooperative farmers in developing countries. With both the private and public sector involved and competing in the organic guarantee system, the need has arisen to coordinate and harmonize these activities. It has become important for the public and private sectors to get a clear understanding of each other's roles and responsibilities and to make sure that organic farming's credibility not only is legally protected, but also is enhanced via a trustworthy guarantee system. The organic guarantee system must contribute constructively as the organic sector grows out of its niche and increasingly conquers mainstream markets.

The initiative for worldwide harmonization of the organic guarantee system came from IFOAM, which in 2002, together with FAO and the UN Committee on Trade and Development (UNCTAD), organized the first international conference focusing on harmonization (Rundgren and Lockeretz, 2002). Subsequently an international task force uniting the public and private sector evolved and started to develop strategies and mechanisms to allow the organic guarantee system to become a tool for the worldwide development of organic marketing and trade (IFOAM, 2004a). This means that the international exchange is not only of organic products, but also of local and regional marketing activities. With participatory stakeholder involvement, IFOAM has recently developed a practical and workable internal control system for smallholder farmers in developing countries. The latest activities concentrate on developing affordable and cost-effective alternative certification programmes addressing the needs of organic farmers who market their products through local and regional food chains and have only a small turnover that cannot justify the high costs of certification.

IFOAM's flexibility and openness to critical reflection is seen in very candid discussions not only of the achievements of the organic guarantee system, but also of the obstacles faced in its development. It is in the best spirit of the founding fathers and mothers that IFOAM not only never intended to rule the world of organic agriculture, but actively avoided ever trying to do so.

9.7 Promoting Organic Agriculture Worldwide, from the Bottom Up

From its very first days IFOAM has accepted the challenge of promoting organic agriculture internationally. However, it started serious advocacy and lobbying activities only by the late 1980s. IFOAM's entry into the world of international lobbying was triggered by the circulation of the first draft of the EU's regulation on organic agriculture, which defined organic as 'free of chemical residues'. This not only reflected a complete misunderstanding of the concept of organic farming, but was also a threat to further market development. An IFOAM task force immediately started to influence the drafting process, and ultimately was instrumental in getting the current EU regulation to reflect the innovative concept of defining the quality of organic food not only by the characteristics of the end product, but also by how it was produced.

From this EU-focused lobbying, the next step was taken on the occasion of the UN Conference on Environment and Development in Rio de Janeiro in 1992. There the foundation was laid for getting the FAO first to realize that organic agriculture is 'out there', and then to appreciate the potential value of close cooperation between IFOAM and the FAO. Subsequently, IFOAM was granted observer and liaison status with the UN itself and with its relevant components, including the FAO, UNCTAD, the International Labor Organization (ILO) and the UN Environmental Programme (UNEP).

The desire for cooperation with the FAO has been a special challenge in persistence. Not only did it take about 4 years to get NGO observer status, it took a few years more until the FAO officially put organic agriculture on its agenda. Yet once this happened, significant progress was made. Today organic agriculture is among the FAO's five priority areas, and IFOAM's cooperation with the FAO is well established and coordinated through an interdisciplinary working group on organic agriculture. Joint conferences such as the one already mentioned focusing on harmonization, and the First International Organic Seed Conference (IFOAM, 2004b), have become very important elements of working together. Similar cooperation has been developing with the UNEP, especially around biodiversity. This got a boost with the Third IFOAM Conference on Organic Agriculture and Biodiversity, organized with the UNEP and held at its headquarters in Nairobi in 2004.

Looking historically at the development of IFOAM's advocacy activities, it is interesting that international institutions that originally were identified as targets of lobbying have become or are becoming cooperating partners. But IFOAM does not focus its lobbying and liaison activities only on international and governmental agencies. At least as important is its cooperation with like-minded international NGOs. IFOAM's active involvement and participation in the World Conservation Union has had a special impact on the environmental and nature conservation movements. For many years IFOAM has been a driving force to put organic agriculture and biodiversity on the agenda of this world umbrella for the environmental sector. Other 'world player' NGOs with which IFOAM cooperates are Greenpeace, Slow Food and to a lesser extent, the WWF.

Permanent information exchange, common projects and other creative ways of cooperation have also been established over the years with social movements, especially the fair trade movement. Meetings, joint projects and a common network (ISEAL) with the Fair Trade Labelling Organizations International, the Forest Stewardship Council, the Social Accountability Initiative, the Rain Forest Alliance and the Marine Stewardship Council, help the organic and fair trade movements grow together.

Starting with the Fifth IFOAM Trade Conference in Italy in 1999, the organic movement reached out to the emerging movement of food culture popularized and represented by the organization Slow Food. This cultural bridge between farming and food is the latest example of the potential power of synergies through cooperation.

In the rapidly growing area of lobbying and advocacy activities, in particular, it can be seen that IFOAM is not narrowly focused on developing organic agriculture. Rather, it seeks to link with other important movements and sectors of society and inspire them to join in the paradigm shift needed to make our fascinating world even more fascinating and better.

9.8 Ready for the Future

After more than three decades the International Federation of Organic Agriculture Movements is implementing its newly formulated (but somehow always existing) mission to 'lead, unite and assist the organic movement in its full diversity'. Its goal is 'the worldwide adoption of ecologically, socially and economically sound systems that are based on the principles of organic agriculture'. Several objectives have been formulated to make sure that this central goal is achieved. The mission will continue to ensure that the organic movement remains current and up to date by offering platforms for innovative and intensive discussions and consultations. An important recent activity via a worldwide discussion and stakeholder consultation was the fundamental revision of the 'Principles of Organic Agriculture' (see Chapter 8, this volume).

In all, IFOAM seems well positioned to continue dynamically to grow organically as it moves towards its 50th anniversary in 2022.

IFOAM Anthem

(melody: Auld Lang Syne)

In all the world the need is felt
To make a drastic change
A choice for life, a choice for health,
Ever wider is the range.
So let us sing to living soil,
Organic farmers' pride,

IFOAM brings us all together
To reach this goal worldwide.
They herd the cows, they plant the seeds.
Not only humans do they feed.
Also water, soil and air.
So let us sing to living soil
Organic farmers' pride,
IFOAM brings us together
To reach this goal worldwide.
May all our children and their children
Live on a greener earth,
For they inherit all our deeds,
That is what makes it worth.
So let us sing to living soil,
Organic farmers' pride,
Join hands and may the work be blessed
To reach this goal worldwide.

References

Besson, J.-M. and Vogtmann, H. (eds) (1978) *Towards a Sustainable Agriculture*. Verlag Wirz AG, Aarau, Switzerland.

Bourgeois, D. (1997) How it all began. *Ecology and Farming* 17, 12–14.

IFOAM (2002) *IFOAM Norms – IFOAM Basic Standards – IFOAM Accreditation Criteria*. IFOAM, Tholey-Theley, Germany.

IFOAM (2004a) *Harmonization and Equivalence in Organic Agriculture*, Vol. 1. IFOAM, Bonn, Germany.

IFOAM (2004b) *Proceedings of the First World Conference on Organic Seed*. IFOAM, Bonn, Germany.

Rundgren, G. and Lockeretz, W. (eds) (2002) *Reader, IFOAM Conference on Organic Guarantee Systems: International Harmonisation and Equivalence in Organic Agriculture, 17–19 February 2002, Nuremberg, Germany*. IFOAM, Tholey-Theley, Germany.

The Soil Association

P. Conford[1] and P. Holden[2]

[1]88 St. Pancras, Chichester, West Sussex, PO19 7LR, UK;
[2]Director, Soil Association, South Plaza, Marlborough Street, Bristol BS1 3NX, UK

On May 3rd, 1946, the Soil Association was formally registered, and on May 3rd, 1956, it held its tenth birthday party, having in the interim grown from a few hundred to over four thousand members. The party took place in the hall of the Royal Empire Society, London, and lasted from 3.15 p.m. until late in the evening. Members and their friends were received by Lady Eve Balfour and Lord Newport, and shared in a birthday cake, complete with candles and a compost box modelled in sugar.

(Soil Association, 1956, p. 575)

10.1 The Association's Origins and Founding

The Soil Association – an educational and research body with charitable status, and a limited company – held its inaugural meeting in London on 30 May 1946 with Lady Eve Balfour as the chair (Fig. 10.1). The meeting elected a Council and appointed Lord Teviot as president. The Association's objects were subsequently formulated as follows:

> 1. To bring together all those working for a fuller understanding of the vital relationships between soil, plant, animal and man.
> 2. To initiate, co-ordinate and assist research in this field.
> 3. To collect and distribute the knowledge gained so as to create a body of informed public opinion.
>
> *(Soil Association,* 1947, p. 1)

©CAB International 2007. *Organic Farming: an International History*
(Lockeretz)

Fig. 10.1. Portrait of Lady Eve Balfour by Mary Eastman. (Courtesy of the Soil Association. With permission.)

The immediate impetus to the founding of the Association was provided by the success of Balfour's book *The Living Soil* (1943), which was into its eighth edition by 1948. However, this success was itself the product of an interest in the relationship between soil fertility, methods of cultivation, diet and health, which had been growing steadily for about 10 years. Philip Conford's book *The Origins of the Organic Movement* (2001, pp. 47–64, 81–97) gives a full account of these developments, but certain main features need to be noted here to understand the Soil Association's significance.

The Association emphasized the (literally) vital importance of the biology of the soil, as against those – primarily research stations and companies promoting agricultural fertilizers – who concentrated on its chemistry. Concern about the fertility of British soils was increasingly expressed by the organic pioneers between the two world wars as chemical, or artificial, fertilizers were more insistently advocated, their use increasing dramatically from 1940 onwards. In India, the English botanist Albert (later Sir Albert) Howard had demonstrated during the

1920s that a biological approach to the soil, based on scientific composting of wastes, substantially improved soil quality and increased disease resistance in crops and the animals that ate them. Howard became one of the Soil Association's chief begetters.

Another central figure, who also spent many years in India, was the nutritionist Robert (later Sir Robert) McCarrison, whose study of the remarkably resilient Hunza tribesmen on the north-west frontier led him to conclude that methods of cultivation might affect human health: the Hunzas practiced composting and their whole food diet was a model of nutritional wisdom. Might their example be relevant to the poor health and physique so widespread among the British? From the mid-1920s, McCarrison was in touch with the doctors George Scott Williamson and Innes Pearse, who ran a family health club in south London that grew into a major experiment in social medicine, the Pioneer Health Centre. All three were convinced that health was a positive, dynamic state of vitality that could be created by improving environmental conditions, diet being prime among them (Conford, 2001, pp. 130–145).

By the mid-1930s, the hypothesis was being developed that human and animal health would benefit from food produced by a 'living' soil, rich with the humus created by obeying the 'rule of return' of biological wastes. As a corollary to this, the early organicists suggested that reliance on chemical fertilizers might adversely affect health, in ways not yet apparent.

These ideas were taken up by the agriculturalist Viscount Lymington (1938), whose book *Famine in England* powerfully affected Balfour. She had farmed at Haughley in Suffolk since 1919, and felt the implications of Lymington's book to be of major national importance, his themes requiring experimental examination. War delayed the start of what became known as the Haughley experiment, but Balfour drew together the strands of the organic case in *The Living Soil*, written to raise money for the experiment. The book looked ahead to post-war reconstruction, arguing that an agriculture based on organic principles might prove a valuable form of preventive medicine.

The number of letters that *The Living Soil* generated seemed to demand the establishment of a clearing house for exchanging information, and Balfour, Scott Williamson and the Wiltshire farmer Friend Sykes were the leading spirits in creating a body that became the Soil Association. They organized a founders' meeting held on 12 June 1945, inviting figures who might be generous with their time and money; after various meetings of the Founders' Committee, the inaugural meeting took place in May 1946. (Details of these meetings and a list of those who attended the founders' meetings can be found in the Soil Association minute book, at the Association's offices.)

Howard attended the founders' meeting but later refused to be involved, apparently objecting to certain aspects of the Association's scientific policy. Nevertheless, the 109 founder members included many notable figures from various fields, among them doctors, dentists, farmers, journalists, engineers and horticulturalists. The Association was politically non-partisan, although at that time its roots were stronger on the Right than on the Left. Its early members were united by an implicitly, and often explicitly, religious philosophy of faith in 'Nature's fixed biological laws', which had to be obeyed if humanity was not to create its own nemesis. 'There are no materialists in the Soil Association', Balfour is reported to have said. (The political and religious affiliations of the early organic movement are discussed in Conford, 2001, pp. 146–163, 190–209.)

The name of the Association's journal, *Mother Earth*, was suggested by Scott Williamson, who believed that the phrase accurately captured humanity's dependence on the soil for its existence and sustenance. After two introductory, explanatory issues, *Mother Earth* appeared quarterly from spring 1947 onwards, with Jorian Jenks as editor until his death in 1963. As well as launching its journal, the Association during its first year began to organize an advisory service and increased its membership to over 1000.

10.2 From 1946 to the Early 1970s

Despite significant public interest in the organic case during the war, the Soil Association was going completely against the trend of post-war agricultural and medical policy. The 1947 Agriculture Act confirmed Britain on the path of intensive chemical and mechanical methods, while the National Health Service, established the following year, emphasized cure rather than preventive measures and paid minimal attention to diet. But the impact made by Rachel Carson's *Silent Spring* (published in Britain in 1963) and the subsequent appearance of environmentalist classics like *The Ecologist*'s tract *Blueprint for Survival* (1972) and E.F. Schumacher's *Small is Beautiful* (1973) finally lent some weight to organic ideas. To describe the Soil Association's work during this quarter century, it will be helpful to consider it under three headings: the Haughley experiment, *Mother Earth* and various forms of promotional and educational activity.

10.2.1 The Haughley experiment

Balfour's aim was to 'inquire into the importance of the vital attributes of the soil for the health of plant, animal and man' (1976, p. 187), and the experiment's strategy involved farming three comparable areas by three different methods. Two plots carried stock, one being cultivated organically and the other with a mixture of wastes and artificial fertilizers. The third was stockless, relying on artificial fertilizers and crop residues.

The Soil Association took responsibility for the experiment, an involvement that continued until the experiment was wound up in 1969. A scientific management committee appointed a board of independent observers, and in 1952 the biochemist Reginald Milton began a 12-year stint of sampling and analytical work. In 1957, the journal *Nature* (3 May 1957, p. 514) praised the unique importance of the long-term ecological research undertaken at Haughley. An account of the results can be found in the updated edition of *The Living Soil*, but the final verdict must be that the experiment failed to achieve its stated aim of tracing the relationship between soil treatment and health, the variables being far too numerous and complex (Balfour, 1976, pp. 211–365). There were some smaller-scale findings that appeared suggestive: for instance, it was discovered that humus content tended to rise on the organic section and decline on the stockless section (the scientific significance of the experiment is discussed further in Chapter 4, this volume).

Lack of funding was a perpetual problem, and a financial crisis in 1966 was resolved only when the property developer Jack Pye acquired the freehold of the farm and backed the experiment with generous grants for 3 years. However, following dissension on the Association's Council, it was decided to end the experiment and concentrate on demonstrating that organic farming could be both productive and financially viable.

10.2.2 *Mother Earth:* the Journal of the Soil Association

The Soil Association has attracted criticism for being 'unscientific' or 'mystical', but its journal was in fact strongly scientific during its first three decades, publishing specialized articles on soil and water science, composting, forestry, biological pest control and the development of form in evolution, among other topics. Another notable feature was an emphasis on ecology, years before *Silent Spring*.

The journal argued consistently that the Association was opposed not to science as such, but to reductionism, being committed to a holistic approach instead: one editorial described the Association's central ideas as 'applied ecology' (Soil Association, 1952, p. 3). Some prominent scientists in sympathy with this philosophy supported the Association, most notably the naturalist Frank Fraser Darling, the director of Monks Wood Experimental Station Kenneth Mellanby, agriculturalist Professor Lindsay Robb and the French grassland expert André Voisin.

Mother Earth alerted readers to issues that later became matters of public concern: factory farming, antibiotics in agriculture, food additives, fluoridation, the impact of industrial farming on wildlife and landscape and the effects of a declining agricultural workforce on rural economy and culture. One issue frequently covered was the technology of municipal composting. There were many articles of practical value to farmers and gardeners, and developments in agriculture at home and abroad were closely monitored.

Under Robert Waller, who edited the journal from 1964 until 1972, there was a shift towards discussion of broader cultural issues and an increasing emphasis on environmentalism. Grist for the mills of those eager to condemn the Association's 'mysticism' was the persistent presence in the journal's pages of those who stressed the spiritual philosophy implicit in an ecological vision of Nature's laws.

10.2.3 Promotional and educational activities

By the mid-1950s, the Association's membership was around 3500, from more than 50 countries. Balfour was a major reason for the growth of overseas membership, undertaking three major tours of North America during the 1950s and also visiting Australia, Italy and Scandinavia. She made regular summer tours of the British Isles, encouraging the work of local branches.

The Association exhibited regularly at the Chelsea Flower Show and at agricultural shows; displays at the Royal Agricultural Show included exhibition plots. There was always sufficient interest in the organic philosophy to ensure that the Association's mail order department was busy dispatching reprints of journal articles and books on organic growing and nutrition.

In 1960, the Association stepped into the commercial world when the Wholefood shop was opened in London, an initiative aimed at providing an outlet for the produce of farmer and grower members and a source of organic food for consumers. The preparatory work was undertaken by the Association's general secretary, C. Donald Wilson.

Five years later, the Association was midwife to another nutritional organization, when a group of doctor and dentist members at the annual conference decided to form a society devoted to dietary studies; they named it the McCarrison Society in tribute to Sir Robert.

Doris Grant, a long-term member, promoted the Association's ideas on nutrition in her cookery books, such as *Your Daily Food* (1973). Another member, Ruth Harrison, wrote the influential polemic against factory farming, *Animal Machines* (1964). From time to time the Association produced educational films to communicate its ecological philosophy: *The Cycle of Life* (1950) was an impressive study of the 'rule of return', while in 1969 *The Secret Highway* rather overambitiously attempted to demonstrate the damaging environmental effects of modern farming (copies of the two films exist on videotape).

Following the abandonment of the Haughley experiment, the early 1970s was a transitional period for the Association. The Earl of Bradford, president since 1951, was replaced in 1971 by Schumacher, who soon became the guru of environmentalism. E.F. Schumacher, Balfour and the new general secretary Bill Vickers formed an Action Group to guide the Association through its difficulties, and in a speech given in June 1971 Schumacher outlined his ideas on how it should develop (Schumacher [1971] is an edited version of his speech).

He urged the Association to demonstrate that organic farming was viable, and to gather and communicate all available evidence on the subject of wholeness. It was time to incarnate the Association's values in the world of commercial interests and to challenge the ideas on which that world was based.

10.3 From the Early 1970s to the Mid-1980s

Schumacher's speech, delivered to mark the 25th anniversary of the Soil Association, proved prophetic and makes fascinating reading more than 30 years later, both because of its inspirational content and because so much of what he predicted actually came to pass in the last quarter of the 20th century. However, the challenges he outlined in his address clearly were unlikely to be realized without a considerable struggle. This is certainly reflected in the history of the Soil Association's development during the 1970s and 1980s. At the time of his address, the Council still consisted of a mixture of long-standing and, in some cases, founder members, fiercely loyal to the founding principles and more than a little nervous about the idea that the organic philosophy could be commercialized and find expression in the marketplace.

A clash between generations became evident during the 1970s. One of the Association's most active local groups, based at Epsom in Surrey, ran a series of successful summer courses on biological husbandry during the decade, under the guidance of Anthony Deavin, at which it became evident that there was a certain cultural divide between some of the Association's officials and the predominantly young students who attended. Similarly, a veteran of the whole food movement recalls that while he found the Soil Association a valuable source of information on nutrition and agriculture, he saw it as essentially a rather establishment-based club for farmers (Anthony Deavin, London, April 2005, personal communication; Gregory Sams, Epsom, November 2005, personal communication).

The Association had since its founding emphasized the importance of an ecological perspective, but was not benefiting from the surge in environmental concern that marked the early 1970s, even though the president's book *Small is Beautiful* was a bible for environmentalists.

The Association's membership remained stubbornly at around 4500 or less, and financial constraints meant that the journal (by now simply called the *Journal of the Soil Association*) was austerely produced, lacking both the depth of *Mother Earth* and the visual appeal that might have attracted younger readers.

The 1970s was therefore a period of struggle, but it was exactly during this time that a new force emerged in the organic movement: a group of young farmers and growers, many of whom had grown up in an urban environment and nearly all of whom had been influenced by the culture of the late 1960s and early 1970s, with its mistrust of urban, corporate life and the industrialism that underpinned it. These new entrants lacked nothing in enthusiasm, but despite being drawn into the organic ideal had no detailed knowledge of the history of the Soil Association. They were untrained in conventional agriculture and in many cases were quite literally learning how to farm organically through practice. In some cases, the influence of self-sufficiency guru John Seymour (Seymour, 1973), himself a Soil Association member, was a factor. Many of them settled in the West Country and Wales, where property was cheap (David Frost, Peter Segger and Carolyn Wacher, Aberystwyth, June 2005, personal communications).

The one feature they all shared was a determination to make a living using organic methods. However, they quickly discovered that the prevailing economic climate, dominated by the European Union's (EU's) Common Agricultural Policy, was extremely hostile to low-input and sustainable production systems. They concluded that the best way to remain financially viable would be to take their products to the market and obtain premium prices from consumers who shared their 'green' beliefs and ideals.

The development of organic marketing necessitated the creation of standards to protect consumers through defining and policing the integrity of the production system. It also meant that producers had to be educated, organized and, when required, disciplined. Early drafts of Soil Association standards dating back to 1967 were developed into the first standards document, published in 1973. The process of refinement and development of these standards has been continuing ever since.

By the mid-1970s, this new group of producers had become involved in the affairs of the Soil Association. Peter Segger was elected to the Council in 1975, and four others – Francis Blake, Patrick Holden, Ginny Mayall and Carolyn Wacher – joined in 1980 (Fig. 10.2). The old guard of the Council saw the arrival of these 'Young Turks' as both a threat and a challenge. There followed a period of great tension and acrimonious debate culminating in the resignation of the then president, Lord O'Hagan, and Lady Eve Balfour, after a stormy Annual General Meeting in Edinburgh in 1982.

Fig. 10.2. Some prominent figures in the British organic movement (primarily) in the 1980s, including several mentioned in this chapter. (Collage by Nick Rebbeck and Carolyn Wacher. With permission.)

But the arrival of this new generation of organic farmers and growers was not entirely rejected by the old guard. In particular, Mary Langman, long-time friend of Balfour, partner in Wholefood of Baker Street, and founder member of the International Federation of Organic Agriculture Movements (IFOAM) and of the Soil Association's Organic Standards Committee, quickly realized that the ideas of the incomers did not actually represent any challenge to the Association's founding philosophy. Rather, they were a manifestation of what Schumacher himself had argued was necessary in his 1971 speech. As a direct result of Mary Langman's influence, particularly on Lady Eve, a damaging and possibly irreversible split in the Association's development was averted (more about Mary Langman may be found in Woodward, 2004). This made possible an ongoing relationship between the new group of younger Council members and Lady Eve herself until her death in 1990, shortly after she had been awarded an OBE.

By the early 1980s, the centre of gravity in terms of influence and policy had already shifted towards the younger organic farmers and growers. Two producer groups – the Organic Growers Association and British Organic Farmers – were formed in 1981 and 1983, respectively. The formation of these groups had the blessing and support of the Soil Association Council and was instrumental in bringing producers together, creating formal and informal networks, organizing farm walks, events and conferences, as well as helping farmers and growers share technical knowledge and develop their capacity to get their products to the market (Richard Young, Broadway, April 2005, personal communication). These groups founded the magazine *New Farmer and Grower* in summer 1983.

Collectively, these producer initiatives and the development of standards created the platform on which the UK organic market could be built. During this period the Soil Association also started honing its skills as an organization with an ability to communicate with the public through the media. In this way the embryonic but increasingly productive capacity of the organic community was investing in cultivating awareness among its consumer counterparts, thereby creating the means for a major expansion of the organic market.

In the early 1980s, the Association's quarterly journal was given a more contemporary format and the range of its contents was expanded considerably to include items on campaigns, global environmental issues, relevant press items, member profiles, news from members' farms, recipes and detailed interviews and book reviews. In 1984, west Wales members Patrick Holden and Peter Segger aroused much interest when interviewed on BBC Radio's 'On Your Farm'. The Association was attacking government food strategy, campaigning against spray drift and pesticides and, long before the bovine spongiform encephalopathy (BSE) scare, had banned animal protein in ruminant rations.

Its higher profile and more combative approach meant that membership began at last to start climbing, albeit slowly at first.

The following year, and in large part owing to the influence of the Wales and West Country groups, the Association moved its office from Haughley to Bristol. Balfour had started farming at Haughley 66 years earlier, so this was a significant break with the past. But it ushered in a period during which the Association would make impressive headway.

10.4 From the Mid-1980s to the 21st Century

Within 15 years of Schumacher's presidential address, the organic food market had developed to the point where annual UK turnover of organic produce exceeded £5 million, and the major supermarkets were in discussion with the Soil Association about supplies. Applications from producers to become Association symbol holders were almost more than the Association could cope with at this time. Holden, Segger and Lawrence Woodward had established links with officials at the Ministry of Agriculture, Fisheries and Food, and when the UK Register of Organic Food Standards (created in 1987) published its standards in 1989, they closely followed those of the Soil Association.

From the end of the 1980s, a succession of food scares in the UK began to undermine consumer confidence in agricultural orthodoxy. These included concerns about residues of pesticides, antibiotics and hormones in food, salmonella and other forms of bacterial contamination, BSE and finally genetic engineering (North, 2001). These food scares eroded the public's trust and confidence in intensive farming and whetted their appetite for the organic alternative.

By the end of the 1980s, a considerable investment had been made in cultivating the interest of British supermarkets, and they all entered the organic marketplace. Safeway was the first to stock organic products, in 1981, followed by Waitrose, Sainsbury, Tesco and Marks & Spencers, all by the end of the decade. As a result, the supermarkets were well placed to respond to the consumer interest created by the food scares.

Parallel with the expansion of the organic market, the Soil Association developed as the principal support, communication and certification organization representing the whole organic food chain under one organizational roof, and was able to press the organic case. As early as 1993, it banned the presence of genetically modified organisms (GMOs) in organic foods and, the same year, launched a project to foster local food links. The Association's staff increased from 9 in 1990 to 140 by the turn of the 21st century, and its income from £326,000 to £5,125,000 during the same period.

The Association campaigned on various issues during the 1990s, including reform of the Common Agricultural Policy, the banning of organophosphate pesticides and the need for government targets for the increased production of organic foods. In 2000, its biodiversity report provided evidence that organic farming encourages wildlife, and it has continued to fight an effective battle against GM crops. In 2001, the Association's chair, Helen Browning, was appointed a member of the Future of Farming Commission, which issued the 'Curry Report' (officially called *Farming and Food: A Sustainable Future*) in 2002.

10.5 The Soil Association Today

Early in the 21st century, the Soil Association finds itself on the threshold of perhaps the biggest challenge it has confronted in the whole of its 60-year history. The combined forces of the industrialization of agriculture, globalization and the loss of loyalty and trust of the consuming public in the farming business have collectively reduced the status of farmers to mere commodity producers whose raw materials are often sold at less than the cost of production (Harvey, 1997; O'Hagan, 2001; Benson, 2005). To address these problems, the Association has concluded that a major and sustained investment in rebuilding public awareness of the importance of farming, starting in schools and achieved through parallel growth in its membership and support base, can avert the catastrophic consequences of the urbanization of the farming community. This 'flight from the land' is taking place all over the planet, but has already reduced the percentage of those involved in British agriculture to around 1% of the workforce.

Organic produce is now big business: in 2006 annual UK sales exceeded £1.6 billion. The Soil Association's membership, which did not pass 5000 until the late 1980s, had reached 17,270, plus 4766 symbol holders, by November 2005. Success has, of course, brought problems. There are those who feel that the close links between the organic movement and the supermarket chains may not be entirely compatible with the Soil Association's wider ecological and social vision. Some Association members – perhaps predominantly, although not exclusively, older ones – fear that its philosophical basis has been forgotten or eroded. Whereas *Mother Earth* contained both science and philosophy, the Association's current magazines present a somewhat less holistic picture, with *Living Earth* tending to concentrate on consumer issues and *Organic Farming* providing specialized information for farmers and growers. The spiritual philosophy has largely disappeared from both.

Nevertheless, the holistic outlook of the Association's founders continues to be embodied in the wide variety of activities it undertakes, and in the range of its concerns, from local farmers' markets to world fair trade. Its organizational structure, involving all the links in the so-called food chain, including producers, food processors, retailers and the public, and its direct involvement with standard-setting and certification, leave it in a strong position to influence the development of organic agriculture in the UK. It has an impressive inheritance of theory and practice that it can share with other organizations throughout the world to ensure that the powerful ideas that motivated the organic movement's founders in the middle of the 20th century can fulfil their potential to bring about lasting change in the 21st century.

10.6 The Future of the Organic Movement in Britain

Movements pass through cycles, and the organic movement in Britain is now reaching the end of the second phase of its 60-year history. From the founding of the Soil Association in 1946 until the mid-1970s the emphasis was on setting down the basic philosophy and principles. During the past three decades the focus has shifted to ensuring that organically grown food is widely available to the public through a variety of outlets, chiefly the supermarkets.

This second phase has been remarkably successful: sales of organic products in Britain have now reached £1.6 billion a year, and continue to increase at a dramatic rate. Success brings its own problems, though, and it is regrettable that British producers are unable to meet the demand from consumers, so that the majority of produce sold has to be imported.

Then there is the question of standards, which have been painstakingly established over a long period. The organic movement's commercial success has inevitably attracted those who do not share its philosophy and who put the search for profit ahead of principles, with the consequence that public trust in the accuracy of labelling may be eroded. In Britain, the movement's honeymoon with the press appears to be over, while EU regulations may pose another threat to the integrity of organic standards. The Soil Association nevertheless is confident that these problems can be tackled.

We must now ensure, as we move into the next phase of the British organic movement, that the recent emphasis on marketing and consumerism does not obscure the new strategic priorities that are emerging. We need to recall that the movement's pioneers saw organic cultivation as part of a wider, ecological vision of humanity's relationship with the natural world, a vision that is more relevant today than ever before.

In 2006, David Miliband, the Secretary of State at the Department for Environment, Food and Rural Affairs, challenged UK farming to achieve 'one-planet' agriculture: farming that works within the environmental and resource limits of Earth. The present system, based on the assumption that cheap oil would be available indefinitely, is unsustainable and unlikely to last more than another 15–20 years, if that. Organic farming is ideally placed to offer a viable alternative.

True, most organic produce in Britain is sold through chains that themselves have depended on the cheap-oil economy, and this has attracted criticism. But the Soil Association has worked hard to encourage farmers' markets and local produce, and the message has even reached the chief organ of mechanized agriculture, *Farmers' Weekly*, which now accepts the necessity for a drastic reduction in the distance that food travels.

The collapse of the present system of food production is a fearful prospect, unless an alternative can be developed soon enough to take its place. The UK organic movement, with its emphasis on healthy food produced by means that respect the environment and minimize the use of non-renewable energy, now has an opportunity to accept the government's challenge and demonstrate that by marrying the principles of its founders to the skills of its present practitioners it can provide 'one-planet' agriculture.

References

Balfour, E.B. (1943) *The Living Soil*. Faber & Faber, London.

Balfour, E.B. (1976) *The Living Soil and the Haughley Experiment*. Universe Books, New York.

Benson, R. (2005) *The Farm: The Story of One Family and the English Countryside*. Hamish Hamilton, London.

Carson, R. (1963) *Silent Spring*. Hamish Hamilton, London.

Conford, P. (2001) *The Origins of the Organic Movement*. Floris Books, Edinburgh.

The Ecologist (1972) *Blueprint for Survival*. Penguin, Harmondsworth, UK.

Grant, D. (1973) *Your Daily Food*. Faber & Faber, London.

Harrison, R. (1964) *Animal Machines*. Vincent Stuart, London.

Harvey, G. (1997) *The Killing of the Countryside*. Cape, London.

Lymington, Viscount (1938) *Famine in England*. Witherby, London.

North, R.A.E. (2001) *The Death of British Agriculture*. Duckworth, London.

O'Hagan, A. (2001) *The End of British Farming*. Profile Books, London.

Schumacher, E.F. (1971) *Journal of the Soil Association*, October, pp. 313–316.

Schumacher, E.F. (1973) *Small is Beautiful*. Blond & Briggs, London.

Seymour, J. (1973) *Self Sufficiency*. Faber & Faber, London.

Soil Association (1947) *Mother Earth*, Harvest.

Soil Association (1952) *Mother Earth*, July.

Soil Association (1956) Mother Earth, July.

Woodward, L. (2004) Obituary for Mary Langman. *Guardian*, 26 April.

11 Ecological Farmers Association and the Success of Swedish Organic Agriculture

I. Källander

President, Ecological Farmers Association, Gäverstad Gård, 61494 Söderköping, Sweden

11.1 Introduction

With 19% of its farmland managed organically in 2005 – just shy of the national goal of 20% – Sweden is a world leader in reorienting agriculture in a more sustainable direction. Such a big change requires several different tools that complement and reinforce each other: a solid foundation of common values; knowledge and experience; engaged and dynamic individuals and organizations; possibilities for consumers to identify organic products in the evermore anonymous food market; forums for strategy building; political lobbying; interested and bold market actors; and deep, broad cooperation and dialogue among the stakeholders in the whole food sector, from consumers to decision makers and from farmers to scientists. The history of organic agriculture in Sweden is one of people and organizations who, using these tools and aware of both the challenges and the potentials of organic farming, have managed to bring about significant change, a development that, of course, is continuing.

11.2 Agriculture in Sweden

Although Sweden is one of the biggest countries in Europe, its farmland amounts to only 2.8 million ha, about 7% of the total land area. The climate allows farming only part of the year, but it is favourable in that the cold winters inhibit infestations of many insect pests and diseases. Agricultural conditions, activities and traditions differ a great deal from north to south. The long summer days in the north make the short growing season very intensive and allow production of high-quality potatoes, berries and vegetables. The northern parts and the forested areas in the south produce milk and meat on grasslands and leys, while grain production is concentrated on the flat and fertile clay soils in the south. Fruit and vegetables are to a great extent located in the south and on the islands of Gotland and Öland.

11.2.1 Structural change

As in many other countries, agriculture has gone through major structural changes in the past half century. In 1961 Sweden had 233,000 agricultural holdings, but by 2005 the number had decreased to 75,000 (Statistiska Centralbyrån, 2006). Between 1990 and 2004, the average farm size increased from 29 to 40 ha. The number of dairy farms and pig farms decreased while the number of animals per farm increased. More diversified medium-sized family farms are disappearing as production shifts to larger, more specialized farm enterprises. Land use has also changed significantly because of globalization and the current price and subsidy systems. The production of legumes, grains and sugarbeets has increased while oilseed production has decreased.

In 1995, Sweden became a member of the European Union (EU) and consequently part of the Common Agriculture Policy (CAP), which substantially influenced Swedish agriculture. Sales became a smaller part of farm income, while direct payments are the most important economic factor for a majority of farms. On the whole, profitability has decreased for all kinds of production, putting agriculture under great economic strain. With the CAP reform in 2005, the conditions for direct payments changed again and many farmers face an even harsher economic situation. Many are looking for alternative enterprises and new markets for their products. For many the economic potential of organic farming is becoming more interesting because of a consistently expanding premium-priced market and the environmental payments to farmers using organic methods.

As everywhere else in Europe, increased use of pesticides and manufactured fertilizers has been harming the environment and food quality. Losses of nitrogen and phosphorus cause nutrient imbalances in lakes, poisonous algal blooms every summer and decreased drinking water quality. Pesticide use has contaminated the Baltic Sea, and decreasing biodiversity in the agricultural landscape and pesticide residues in food are recognized as growing problems. To deal with these problems, national goals and measures for environmental improvement were introduced in the 1980s. With the EU membership in 1995, the Swedish

Parliament introduced a new environmental programme for agriculture with various subgoals, such as preservation of biotopes with particular values, general support for leys and pastures and a programme for organic farming.

11.2.2 The Swedish love of nature

The love of nature – forests, meadows, mountains and grazing animals – is a strong part of Sweden's national mentality, culture and tradition, as expressed in fairytales, songs and traditional celebrations. Author Astrid Lindgren, of Pippi Longstocking fame, and our national poet Evert Taube, whose work is full of beautiful descriptions of plants and animals, countryside life and children's experiences in nature, are among our greatest sources of national pride. Early in school, children learn about flowers and animals, ecology and cyclic systems, and also to respect nature. One of our oldest environmental organizations is called 'Keep Nature Clean', and the Swedish Society for Nature Conservation, with its many regional and local branches, is a powerful institution that fights on behalf of the survival of species, nature and the environment in Sweden and internationally, and has long been the strongest environmental lobbyist. This deep inborn interest for nature and environment is one reason why Sweden is considered among the leaders in environmental development and why organic farming is considered to have an excellent context in which to be recognized as an important part of efforts to improve and preserve nature and environment as a part of a sustainable society.

11.3 Building the Organic Movement

11.3.1 Preparing the ground with concepts and methods

Before the 1980s, organic farming in Sweden was practised and promoted by several organizations, each with its own philosophy, working isolated from each other, often even in conflict. The most important were the Biodynamic Association in Järna, following Rudolf Steiner's teachings, and the Association for Organic Biological Growers, based on the theories of Rusch and Müller (see Chapter 2, this volume). These organizations attracted producers and consumers, and developed growing techniques more than anything else. Many courses, study circles and field days were organized throughout Sweden, and the number of interested producers grew quickly. Consumers were often linked to growers in local/regional groups, and the products, still on a very small scale, were sold directly through local markets, farm shops and box distribution. Through their democratic membership procedures the organizations created a foundation of values and goals that attracted more and more people, both farmers and non-farmers.

11.3.2 Common ground made lobbying possible

Even in the early 1980s, interest from the outside world demanded better cooperation among the active organizations. Five organizations formed a forum of cooperation in 1981/82 to define their common ground under the label 'Alternative Agriculture' and to write a political agenda. This is when organic farming in Sweden became a political movement, something that strongly influenced what happened in the next 25 years and that remains important. Samarbetsgruppen för Alternativ Odling (SAO), the cooperation group for alternative agriculture, brought together farmers, scientists and representatives from the growing environmental movement to formulate a political action programme. Although revised and developed through the years, this programme has remained a very important tool for the continued development of alternative agriculture. Largely because of its strong individual founders, the SAO soon became stronger and more influential in agricultural policy than the organizations that formed it. It discussed alternative agriculture with ministers, organized agriculture policy seminars, produced materials and put on several big exhibits at Sweden's largest agricultural fair.

11.3.3 Need for an organic farmers' organization

With a growing number of 'real' farmers, large-scale as well as small-scale producers, joining the sprouting organic movement, it became more urgent to work with the farmers' main interests and needs, something that the existing organizations did not do well enough. The growing organic output needed a bigger market, and consumers urged their food chains to provide organic food in ordinary supermarkets and grocery stores. A major food chain, the Consumers' Cooperative, responded to this situation but demanded trustworthy control. Knowledge building was increasingly important, and for the professional farmers it became crucial – and possible – to earn a place in national agricultural policy.

That is what happened during a workshop on marketing late in 1984, where a group of the most active organic farmers met to discuss common packaging (see Fig. 11.1). But instead of discussing what they

Fig. 11.1. Gunnar Rundgren (left) and Staffan Ahrén, participants at the 1984 meeting leading to the founding of Alternativodlarnas Riksförbund (later Ekologiska Lantbrukarna, the Ecological Farmers Association). Ahrén became its first president, while Rundgren became the first president of its sister organization KRAV (and later president of IFOAM). (Photo by Inger Källander.)

were supposed to, a new discussion took place. It was about the need to create an organic farmers' association.

The handful of farmers who got together to shape the new organic farmers' union were young, enthusiastic, pioneer organic farmers, most with a political engagement in environmental issues and an academic background behind them. These activists showed themselves to be not only creative and successful farmers, but also skilful organizers. The ground was prepared, the future tasks were clear, statutes and policies were formulated, and in a few months the Alternativodlarnas Riksförbund (ARF), the National Association of Alternative Growers, held its first general assembly in February 1985.

Strategies and plans were forged at the kitchen tables of a few devoted and hard-working young farmers with the goal of building a strong popular movement. Important tasks were distributed among the first eight-member board. The first urgent issue, already planned and negotiated with the Consumers' Cooperative, was the founding of KRAV, the certification body. (KRAV originally stood for *Kontrollföreningen för alternative odling*, Control Body for Alternative Agriculture. Today, with alternative agriculture renamed 'ecological agriculture',

KRAV remains as the name, but is no longer an acronym.) A few board members took on the task of organizing the market, with others developing agricultural policy and lobbying. An effort to enrol as many organic farmers as possible yielded more than 100 members in the first year, organized in active regional groups. By 2005, when it celebrated its 20th anniversary, its name in the mean time having become Ekologiska Lantbrukarna (Ecological Farmers Association, EFA), it had 3000 members organized in 23 regional sub-organizations, allowing organic farmers all over Sweden to be directly active in a very dynamic organization's activities and policy building.

11.4 The Ecological Farmers Association

Farmers' organizations everywhere have always played an important role in the development of organic farming. Organic farmers have been involved in developing practical production systems and methods, working on social issues affecting farmers and rural populations, developing concepts and lobbying for them and to a great extent developing markets. This has given organic farming a solid platform that integrates visions with practical realism. Farmers have been the main driving force in finding creative, practical solutions that advance the goals of organic agriculture. Along with trustworthy certification systems, farmers' constant efforts to improve organic production are the major reasons that consumers trust organic products. Therefore, the main goals of the EFA include maintaining organic integrity and keeping farmers engaged in the continuing development process. Other overall goals are to promote organic agriculture and work for the interests of organic farmers, focusing on initiatives to advance organic farming both quantitatively and qualitatively. Equally important is to define and strengthen the basic values of organic production and to make them known and understood.

The EFA has managed to unite organic farmers in Sweden under one strong umbrella organization. Most members are active farmers, but advisors, teachers and others who want to support organic agriculture are welcome to join. The EFA is a non-profit organization and is unaffiliated religiously and politically. Its main areas of activity include policy and lobbying work, standards development and certification, market development, research, information, and organizational development and networking.

Policy and lobbying: The EFA is constantly involved in lobbying, commenting on proposals from the government and other institutions, and holding discussions with decision makers, and has an expert role in most national agricultural policy work. Since the government started working with national targets for organic agriculture, the EFA has participated actively in the analysis, elaboration and evaluation of various programmes, especially the rural development programme of the EU CAP, which has a special programme for organic production.

Standards development and certification: Through the EFA, which was the initiator and remains an active member of KRAV, organic farmers can influence the standards development and certification process as well as the goals and development policies of KRAV. The EFA also is active in international certification work through the EU group of the International Federation of Organic Agriculture Movements (IFOAM).

Market development: Secure, long-term markets offering premium prices are perhaps the most crucial prerequisite for organic farmers' economic success, more so than government funds. Therefore the EFA increased its efforts to develop the organic market. It decided against setting up a parallel retailing channel, but none the less, there are many ways that farmers as a group can be active in marketing. The strategies that the EFA found worked best are to get market actors to cooperatively analyse bottlenecks and to coordinate initiatives, to provide analysis, information and comments on market development, to develop efficient concepts for information and education of consumers and decision makers on different levels, and to support local initiatives such as farmers' markets, box schemes and cooperative local/regional marketing projects. The EFA also monitors a Market Council made up of organic organizations.

Research: Relevant, adequately funded research programmes are of great importance for the future expansion and sustainable development of organic agriculture and for the success of individual farmers. The EFA is part of the board of the Centre for Sustainable Agriculture at the Swedish University of Agricultural Sciences, and many farmers have contributed their experience to creating a new holistic and system-based framework for organic research. Since 2001, in large part because of successful EFA lobbying, organic research has been granted considerable national funding. The EFA is involved in the evaluation of research and the distribution of research funds at several agencies and institutions. Participatory research is a new field in which organic farmers, extension staff and advisors cooperate.

Information: The EFA's magazine *Ekologiskt lantbruk* (Organic Agriculture) is published 10 times a year, with articles on practical techniques, ideas and policy debates, market and certification information and organizational activities. By focusing on special topics such as genetic engineering and energy use, the magazine prepares the ground for discussions of future policies. Most of the interesting news and debates can be found on the EFA's web site (available at: www.ekolantbruk.se). Several mailing lists have been established to serve various needs within the organization. The EFA publishes a weekly electronic newsletter, as well as books, magazines and leaflets for farmers and students, along with non-technical educational material for non-farmers. A video, a slide show and fact sheets on current issues help the members to educate themselves and others and to participate in the debates over organic farming. Most courses in organic agriculture are organized by the advisors or the agricultural schools, but farmers often participate, contributing their experience.

Organizational development: The EFA's small office is situated in Uppsala, but its staff members are spread out over Sweden (most are also active farmers). The office offers services to members, distributes materials and organizes and facilitates member activities and various arrangements for cooperation within the organization. The EFA also coordinates cooperation within the organic food sector of Sweden. Contact, dialogue and cooperation with other environmental organizations and the Swedish Farmers' Federation, the LRF, are part of the strategy to spread organic ideas. The EFA has created different forums for dialogue according to current needs, and worked very actively for the Ekologiskt Forum, hosted since 2002 by the Royal Academy of Agriculture and Forestry, a highly prestigious private institution (see Section 11.5). As a member of IFOAM and an active participant in the IFOAM EU group and the Nordic cooperation, the EFA supports global efforts to develop organic agriculture with a consensus approach.

A grassroots organization: Many activities, including political lobbying and strategic seminars, take place on the national level through the national board and staff. But to support this important work the members of the EFA are organized in regional chapters to which all current issues are forwarded for discussion. Regular discussions are held on urgent issues, such as the latest policy proposal from the government or the revised standards proposed by KRAV. In this way any member can be active and influence the policies and work of the EFA. The 23 regional bodies of the EFA are responsible for regional activities where the farmers, often with an advisor as coordinator, organize regional courses, field days, harvest parties, research projects, farm days for consumers and other marketing activities, as well as contacts with local decision makers and media. Conventional farmers are actively invited to participate in activities to help them get to know and feel comfortable with organic agriculture and its practitioners. There is often local cooperation between organic farmers and other environmental organizations.

The members can also participate in special commodity-specific e-mail groups for milk, meat, grain, vegetable or egg production. These discussions deepen the general policy and development work.

They also help keep farmers with different interests together on a common platform for organic agriculture and let them speak with a common voice.

11.5 EFA and Agricultural Policy: Targets, Strategies and Milestones

Organic farming in Sweden has had a strong and steady growth since the mid-1980s, especially after Sweden entered the EU in 1995, when the organic area increased from 50,000 to 300,000 ha in 5 years (Rydén, 2003). With its strong consumer and political support and interesting economic features, organic farming has long left behind its image of niche production for a few rich consumers. The old picture of the organic farmer as a long-haired hippie who farms for the lifestyle and for household consumption has since long disappeared, and has been replaced by that of a modern, market-minded agricultural expert who is out front in meeting future challenges of quality and environmentally sound production.

Although the growth has been strongest since the introduction of the EU-financed support programmes, this positive development would not have been so pronounced without the strong platform of the organic movement and some enthusiastic stakeholders in the food sector. The common policy foundation, realistic targets and successful strategies paved the way for the organic movement to have an important role in agricultural politics despite its low numbers and limited economic resources.

The first milestone was definitely the organization of the organic movement under ARF (later renamed EFA) and the elaboration of a policy platform. After 1985, this made it possible for farmers to convert to organic production, to be certified and find a market for the products and not least to be in contact with, learn from, and be inspired by, other organic farmers. The number of organic farmers registered with KRAV grew steadily, as did the amount of organic food on grocery store shelves. Organic farming was becoming a known concept, with its products sought by consumers.

Another milestone came in 1989, with the first national support for conversion to organic farming. Just before national elections, after a few thrilling days of lobbying by the ARF and a threat to stop deliveries of organic products, the social democrat agriculture minister Mats Hellström, eager to win votes for his party, proposed a concrete support programme for organic farmers, a chair at the Swedish University of Agricultural Sciences and three new regional advisors specializing in organic farming. Besides providing a good incentive for expansion, these measures made organic farming politically acceptable, with organic farmers playing an increasing role in national agricultural policy.

The next major event came in 1993, when after a period of reduced interest in environmental issues in Swedish society, the ARF General Assembly decided to reinforce the focus on environment and organic farming by introducing a special target. The slogan '10% in the year 2000' was put forth by the ARF, and a year later was unanimously adopted by the Swedish Parliament. The 10% target triggered an interesting development in which the food and agriculture sector got a common political framework for the growth of organic farming that put responsibility on many more stakeholders than just farmers and consumers. In 1995, the Swedish Board of Agriculture, together with the EFA, KRAV and others, elaborated an action plan for organic farming – *Aktionsplan 2000* – to actively support the goal. In this plan the Swedish Board of Agriculture defined the goal to be 10% of arable land. The government's adoption of the target and plan coincided with entry into the EU, which gave Sweden the possibility of financing support schemes for organic farming, making serious implementation of the plan possible. Other stakeholders in the food sector elaborated their own 10% targets and worked wholeheartedly to achieve them, and since 1995 the market has expanded greatly in both sales volume and range of products.

The most important product of the 10% campaign was society's broad interest in organic development and its full acceptance as a serious market alternative. It also became recognized as an important tool in the development of an eco-cyclic society, with increasing emphasis being given since 1995 to what it can offer for the rural economy, rural residents and the survival of agriculture as a whole. An evaluation of the support programme by the Swedish Board of Agriculture within the 10% action plan indicated that the measures were successful and the goal was more than achieved: 11% in 2000. But basically no support was granted for the work of the organic organizations, and very little for market development. The same evaluation stated that more funded measures for market initiatives would have been beneficial.

Throughout 1999, strategic discussions were held to set new targets. Emphasizing the value of longer-term plans, the EFA's General Assembly approved a new goal: '302010 – 30% organic production by the year 2010'. In parallel, the government asked the Swedish Board of Agriculture to formulate a new 5-year target. After a careful market analysis, late in 1999, the government launched its second overall target: '20% organic production in 2005'. A new idea was to set specific targets for different commodities: 20% of dairy cows; 5% of pigs and meat chickens; 10% of laying hens; 30% of grasslands; 30% of legumes; 10% of sugarbeets; 5% of fruit; and 20% of other crops.

By the end of 2005, close to one-fifth of arable land was managed organically, and the consequences of this kind of support had become clear. The national policy was to finance the whole scheme under the EU rural programme, where the environmental frame and criteria for support offer the possibility of supporting organic farming with a 50% payment from the EU. This choice means that the support came as an environmental payment to farmers who apply the EU standards for organic production, but are not necessarily certified by KRAV and thus not selling products on the organic market. The 'Swedish model', as it has become known in the EU, meant that organic development was driven by two main mechanisms: on one hand, government support via taxpayers for environmental efforts and public goods, and on the other, consumer support through premium prices. Not only EFA, but the whole environmental movement, both within Sweden as well as outside, has seen two-pronged support as an efficient, reasonable and dynamic method for large-scale conversion of agriculture in a sustainable direction. Politically, however, there was a growing scepticism regarding this model, because with only 40% of organic production certified by KRAV and thus able to provide organic products, the approach was not seen as efficient. Voices were raised to change the model.

A recently built structure that has proven to be important and constructive in reaching out to new but important and sometimes sceptical stakeholders is the Ecological Forum at the Royal Academy for Agriculture and Forestry. The forum grew out of the organic movement's criticism of the government's Swedish Board of Agriculture as lacking both the competence and enthusiasm to deal with this fast-growing sector. A proposal was made, first by the Board of Agriculture and later by the Ministry of Agriculture, that the Ecological Forum be founded in the independent Academy. This actually happened in 2002, and the forum has earned a very good reputation for dealing with the most crucial issues in an atmosphere of openness and broad participation. Of course, the EFA has been very active in this development and in the board that plans the work. The forum made a proposal to the government offering to elaborate the third action plan for organic farming, and was given this task in May 2006, a sign of trust and appreciation in the forum and its work model.

Towards the end of the 2000–2005 period, the organic movement via the Ecological Forum organized a broad and open seminar for stakeholders to discuss the needs and potentials for a new target and action plan, as well as possible new models for elaborating it. There was a wide consensus that working with a broadly agreed-upon strategy is a constructive and dynamic way to develop organic faming. Again the Swedish Board of Agriculture thoroughly studied the effects of organic farming on the environment (pesticides, biological diversity, nitrogen leakage) and animal welfare. The report was finished in a year, but the new target was seriously delayed because of the revision of the larger EU rural programme going on at the same time through other political processes. In March 2006, however, Minister of Agriculture Ann-Christine Nykvist announced the new target: '20% certified production in 2010'. A new and interesting addition to the production target was a consumption target: '25% organic consumption in public kitchens by 2010'. A third action plan is currently under elaboration and will be presented in June 2007. The interesting change this time is that the task has been given to the Ecological Forum. This means that a broader stakeholder cooperation is seen as the best way to work, and will give more depth and quality to the result.

The role and influence of the EFA in this whole political development intrigued a researcher at the Swedish University of Agricultural Sciences, Reine Rydén. In his report *Sailing with the Wind – EFA and Agriculture Policy 1985–2000* (Rydén, 2003) he asks what lies behind the successful work of the organic farmers. One factor is the concept of 'policy network', meaning cooperation among a small number of persons with great expertise and competence. The EFA had this competence and was working in a field with bright future possibilities. Rydén comments that the EFA also showed determination by formulating concrete proposals and realistic targets. The success was also, in part, a consequence of a period of good expansion for organic farming in the entire EU. Politicians of all stripes have seen organic agriculture as a positive alternative, something that was evident in the adoption of the 10% target.

11.6 EFA and Certification: Birth and Development of KRAV

KRAV, the Swedish organic certification body, was founded in 1985, only 2 weeks after the ARF. It quickly became a major factor in Sweden's organic development. From the beginning the strategy was to build a control body that was independent, but where farmers and others interested in serious and trustworthy certification could express their opinions and thereby base standards development and certification on mutual understanding and compromise. The four farmer/grower organizations that founded KRAV decided to build neither a system where the farmers control themselves, nor a government institution, but rather an independent organization, privately owned by its members. Besides the four founding organizations, a rapidly growing number of member organizations continued to develop KRAV. One of the first to join KRAV as an active and devoted member was the LRF, the Swedish Farmers, Federation (the conventional farmers' organization), which, although not enthusiastic about organic farming, wanted assurance of a good control system for those who converted to organic farming. Another early member was the Consumers' Cooperative. In 2006, KRAV counted 29 member organizations from the whole food sector: organic and conventional farmers' organizations, environmental and animal rights organizations, the food industry and others.

Standards development for organic agriculture had already started before the birth of KRAV, but now this work could continue in a stronger and very relevant context. There was direct cooperation with the other Nordic countries to work in the same direction, and not long after its founding, the same people who were building KRAV also became closely engaged in the international certification in IFOAM (see Chapter 8, this volume). Several people from the early Swedish organic movement have been very active in different bodies and working groups in IFOAM (e.g. Gunnar Rundgren and Eva Mattsson, who, respectively, have been president of IFOAM's World Board and chair of the Standards Committee); also, KRAV was the first certification body to be IFOAM-accredited.

KRAV has probably been the most important factor in the growth and success of organic farming in Sweden, especially for organic products sold anonymously, such as in supermarkets and the export market. The control and label help the consumer find organic products in their ordinary food stores and guarantee that the farmer has complied with the organic standards. The label also tells the consumer about organic values, and thus is an educational tool. By offering consumers trustworthy control and a label that is easy to recognize, the market immediately took off and has had strong growth ever since. This model of third-party guarantee has obviously been an efficient and culturally relevant tool during this first phase of organic development.

KRAV is still very important in Swedish market development and serves as a meeting place for a number of important stakeholders. The EFA, which, when it was still called ARF, initiated KRAV and organized the structure for broad participation, has of course remained a key player in KRAV and has exerted considerable influence in its development. However, the EFA is only one of KRAV's many member organizations, and in the last few years, with the organic market growing, conflicting interests have become more obvious. The development of the standards has become more market driven, and the more ideologically based members feel that organic values are being diluted. EFA members are often considered 'green fundamentalists', and sometimes feel marginalized in the certification process.

11.7 EFA and Networking: the Swedish Farmers' Federation

There will be no growth and expansion unless conventional farmers get interested in organic faming and dare to have a go at it. Therefore, the

EFA has always been open to dialogue with conventional farmers. Despite this, both individual conventional farmers and their federation, the LRF, remained suspicious and uninterested for many years. But in the late 1990s, the growing economic possibilities, plus the fact that the LRF was trying to improve its environmental profile through different projects, opened the door to organized dialogue between the EFA and LRF. Of symbolic importance was the decision in 1997 by the LRF's then-president Hans Jonsson to go organic on his farm. A year later, the LRF became a member of IFOAM and Hans Jonsson, together with the president of the EFA, Inger Källander, later made a joint presentation at the 13th IFOAM International Scientific Congress about the importance of openness and dialogue in organic development (Källander and Jonsson, 2000). This contributed to a change of attitude among LRF members and helped take away the negative picture that conventional farmers often had about organic farming. Those who wanted to try organic farming did not have to feel that they would be breaking any unwritten rules. Since 1996, the two organizations have had a continuous dialogue on current issues. They do not always agree on the analysis, but this 'constructive tension' feeds energy to the development and sometimes opens doors for common projects.

Networking and stakeholder cooperation was a major strategy from the start, proven to be efficient and helpful for development and growth. Besides the LRF, a main cooperation partner has been the Swedish Society for Nature Conservation (SSNC), which, more than any other organization, almost totally shares the EFA's visions and values, and the two have developed a strong lobbying partnership, but without any formal organizational structures. Together, they have not only targeted agricultural policy makers and processes, but have also been important allies in the sometimes difficult controversies arising in KRAV, and have developed various market development projects.

11.8 Criticism and Backlashes

Success provokes resistance, as the past few years have shown, probably as a natural reaction to the first tremendous growth and public interest in organic farming. Budgets are not unlimited, and when the government and the EU give higher priority to organic farming, something else will get less. The last few years' media debate has been more aggressive and negative, mainly driven by a handful of researchers but supported by some people in high public positions, and also some journalists. The organic movement has had no problems arguing against its critics and defending organics with good scientific evidence, but it has been much more difficult to get the same media attention.

A special case was the famous trial where a small organization of conventional grain producers sued the Consumers' Cooperative over the facts used in a marketing campaign for organic products. The trial went on for a few years and had a strong symbolic value. The crucial issue was whether it is legal to use environmental facts and arguments in marketing, and whether traders may take an active part in the big conversion to a sustainable society. The Consumers' Cooperative won the case in the end, but in the aftermath retailers are much more cautious in their marketing strategies.

11.9 The Way Forward

This chapter has presented a historical description and simple analysis of the processes and events behind the strong development of organic farming, as of 2006. Moreover, it also has emphasized the individual people involved and their visions, strategic thinking and enthusiasm. Development never happens by itself; it always depends on people. However, new events happen and new people constantly emerge. Soon the facts will have changed, and new analyses will be needed. This is what it means to work in a very dynamic sector with lots of potentials and driving forces, a sector that focuses on challenges and possibilities rather than obstacles and problems.

In 2006, everyone involved in organic farming in Sweden is showing enthusiasm, devotion and hope, and the future is looking very bright. A new action plan is incorporating a lot of new ideas and new thinking. A large national 3-year consumer campaign, planned for launch in 2007, is broadly supported by many stakeholders, including producers, officials, as well as market and consumer organizations. In addition, the Swedish Consumer Agency, in accordance with the new organic consumption goal, recently received 1 million Swedish Krona (SEK) (about €110,000) from the government to develop a consumer strategy for organic products, with the aim of supporting a more privately run consumer campaign.

The organic movement is young again as it was in 1984. The main difference today is its vast body of experience and competence and its immense political and consumer support. After a few years of facing tough critical scrutiny and analysis, which often had the aim of proving that organic farming is no better than conventional and not deserving of special political and economical favours, organic farming has every chance to take off again, and another historical chapter written in a few years could make very interesting reading.

References

Källander, I. and Jonsson, H. (2000) Organic and conventional farmers in new dialogue. In: Alföldi, T., Lockeretz, W. and Niggli, U. (eds) *The World Grows Organic*. Proceedings of the 13th IFOAM International Scientific Conference. Basel, 28–31 August 2000. vdf Hochschulverlag AG an der ETH Zürich, Switzerland, pp. 734–735.

Rydén, R. (2003) *Medvindens tid: Ekologiska Lantbrukarna och jordbrukspolitiken 1985–2000*. Centre for Sustainable Agriculture, Swedish University of Agricultural Sciences, Uppsala, Sweden.

Statistiska Centralbyrån (2006) *Jordbruksstatistisk årsbok 2006*. Orebro, Sweden.

12 MAPO and the Argentinian Organic Movement

D. Foguelman

Vice-President, Movimiento Argentino para la Producción Orgánica (MAPO), Sarmiento 1562 6° F, Buenos Aires, Argentina

12.1 Early Agriculture in Argentina

North-western Argentina was inhabited by ancient agricultural civilizations that introduced irrigated crops in the Andean terraces and high plateaux as far back as 3500 BC. In the flats along the large rivers, the southern Guaranis combined shifting cultivation with hunting and gathering. In the Andes, more than 100 indigenous crops were bred over the centuries, with thousands of varieties having been adapted to the different valleys both by local selection and by official stations. As a means of domination, Spanish colonization broke up local agriculture and reoriented it towards European products such as cattle, sheep, goats, horses, wheat and barley. Most of the rich native tradition was lost in Argentina, although it may still be found today in isolated places in Bolivia, Peru and Ecuador.

By the end of the 19th century, Argentina established itself as a provider of foodstuffs for Europe: first wheat and maize and later refrigerated meat, in what was to become the first big modern agro-industry. The production systems always used the best available conventional techniques. By the mid-20th century many elements of the so-called Green Revolution were adopted, including hybrids, fertilizers and pesticides, and an official institutional system dealing with research, development and extension was created to spread the new practices, built on the US model. It is not surprising that Argentina was among the first adopters of genetically modified crops, and that at present it is the world's second leading exporter of soybeans, 95% of which are genetically modified to incorporate resistance to glyphosate.

12.2 Environmentalism and the Green Movement

It is no surprise that in light of this situation, a strong environmental movement was born in the 1980s, with many alternative and diversified purposes, when the end of a cruel dictatorship allowed the revival of civil society organizations. The movement grew with the goal of achieving planetary sustainability through social, economic and technological actions. It must be noted that this movement arose not with the aim of rescuing forgotten local cultures, but rather as a counterpart of international environmental movements such as those of Europe and North America.

This development included the green organic movement. In 1985, two non-governmental organizations (NGOs) connected with organic agriculture were established: the Green Hope Web of School Vegetable Gardens, and the Centre of Organic Agriculture Studies (CENECOS). Both had the support of some professional experts, who were acting as individuals, not in an official capacity. The activities of these groups were intended to spread the advantages of producing and eating organic, ecological or biological food. This was done by means of field manuals, courses and conferences, some specialized shops, and house-to-house delivery of vegetable boxes.

In the early 1990s, the first Argentinian books on organic farming were published, which quickly sold out. An offshoot of CENECOS was ECOAGRO (Eco-Agro, 1992), an NGO specializing in organic agriculture and livestock for the local and export markets. Other NGOs with the same origin focused on spreading town vegetable gardens and on organic pest control. Still another group aimed at increasing the awareness of agroecology in connection with the Latin American movement MAELA. In the early 1990s, there were intense debates about the role of small farmers, the organic farmer's profile and whether the organic movement should promote an orientation towards local or international markets. Each group finally found its own sphere of action and its own sources of funds and at present there are no quarrels among them, but neither is there strong or enduring collaboration.

At the same time, something unusual occurred. The National Institute of Agricultural Technology (in Spanish: INTA), the organization that led the adoption of the 'hardest' production technologies, also harboured a small cluster of technical people who developed a system of training for family, school and community vegetable gardens in the 1990s, called the PROHUERTA Programme (INTA, 1992–2004). The gardens are not strictly organic, as they are not certified; nor do they follow every detail of the standards. However, they produce vegetables naturally, without synthetic chemicals, in school yards, empty city lots, etc. The programme was aimed at poor people, feeding them through nearly 400,000 gardens, mostly suburban, in 3700 Argentinian towns.

They produce 60,000–80,000 t of fresh vegetables per year, with a mean of about 44 kg per participant. The programme organized a national web of promoters and trainers and of vegetable seed production centres, and trained young people in family farm management. Their chief technical people have been invited to advise similar programmes in other Latin American countries. In 2005, INTA upgraded it, as its services were in greater demand.

12.3 Institutionalizing the Organic Movement

Some members of these NGOs collaborated with the secretary of agriculture in working out the official organic standards, and a dual system was developed involving both the secretary of agriculture and private, local certifiers. This system worked conveniently as a double control that in 1995 allowed Argentina to be recognized by the European Union (EU) as an equivalent country (only five countries were so recognized during the 1990s). This system is recommended by the International Federation of Organic Agriculture Movements (IFOAM) as a suitable strategy for new countries entering international markets. At present, Argentina is self-reliant in certification, having four certifying companies recognized in the USA and the EU. Notwithstanding this, official support for organic activities arose more from the initiative of some officials than from a consistent state policy.

The staff of ECOAGRO joined in forming a new NGO in 1996: the Argentine Movement for Organic Production (in Spanish: MAPO), which was aimed at furnishing a formal, central structure for all parties involved in certified organic activity. From the beginning MAPO was associated with IFOAM. Among its present leaders are many initiators of organic activity in Argentina, joined by experts and teachers. Two stages in its activity may be discerned: the first centred on organizing the 12th International IFOAM Scientific Conference, held in Mar del Plata in 1998 (Foguelman and Lockeretz, 1999), while the aim of the present stage is to promote the organic sector within the country for the domestic and export markets.

Of a total of some 1800 organic farmers, about one-third are MAPO associates. As the oldest organic institution, MAPO represents and brings together a wide range of technicians, teachers, researchers, consumers, traders and farmers. At present, its main concerns deal with farming and farmer subjects, their cooperative association and local market development. A compilation of the history of the movement is found in Pais (2002).

Besides organic publicizing and training through formal and informal teaching and publishing of brochures, magazines and books, MAPO reports market information and is a consultant to the government. It typically organizes two Argentinian or Latin American meetings each year on production and marketing topics, and others of local scope. It is supported financially by its associates, by organizing of meetings and by research funding. It has delegates in various Argentinian regions. MAPO promotes the inclusion of organic subjects in university curricula, enters into cooperative agreements with universities to support organic research by connecting farmers and researchers to do field research, joins official and private groups dealing with sustainable development, receives scholarship students from abroad and is a site for research studies. A project of technical cooperation was set up to back the development of organic production among small peasants, funded by the Italian Institute for Foreign Cooperation, while another EU project focused on local market development. Since 1997, MAPO has had two delegates on the IFOAM's World Board, including the present Vice-President. A MAPO associate represents Argentina on the Standards Committee.

Among MAPO's main goals are the advancement of small and medium farmers and the development of local markets. These are very difficult tasks in a country whose economic structure pushes the centralization of activities and industrial farming, so that the peasants (*campesinos*) have almost disappeared. In other words, in a country based on agriculture there is no longer a place for small farmers, with the rural population now just 8% of the total. In 1998, 72% of certified organic farms were very small; they had only 5% of the organic land and did not hire any extra labour (Foguelman and Montenegro, 1999). Recent estimates indicate that the number of organic farmers has decreased for farms of all sizes, but especially among small tenants. The national experience with small farmers' associations is not very encouraging, although cooperation for advising and certification within organic groups is mentioned by rural sociologists as a good example to follow. Recent initiatives with farmer groups, however, are proving more successful. Many think that organic production is the only alternative capable of maintaining small farmers on their land.

Recently, some new NGOs connected with the organic sector have emerged. One is a commercial lobbying organization made up of the biggest farmers and exporters (Cámera Argentina de Productores Orgánicos Certificados); another is an NGO that brings together the certifying companies (Cámera de Certificadores). Many small, local NGOs are developing. MAPO establishes agreements with them to achieve particular goals, such as by giving support in requests for international funds, backing demands for support from local governments and offering technical training.

It has been active in the question of contamination by genetically modified organisms (GMOs). In 2001, MAPO filed a suit seeking to protect farmers against such contamination and to free them from having to set up buffer zones. The suit requested that the buffers be put in the conventional fields where GMOs were planted, rather than in the organic ones. With this in mind, a ban on transgenic Bt maize was sought, the first of its kind in the world.

The suit presented many difficult aspects, given the unconditional official support for GMOs and the problems in verifying the source of GMO contamination when an organic farm is surrounded by a sea of GMO crops. Hence the suit was dropped for lack of concrete cases to present. The repeated need for analysis of seeds and crops for GMOs makes organic production more expensive, and soybean consignments have already been rejected for exceeding allowed GMO limits. The risk of contamination also strongly discourages honey production, hindering bee-keepers from taking advantage of high demand and good international prices. A further threat is the risk of contamination of maize with Bt-transgenic pollen. Because north-western Argentina is a centre of native maize germplasm, a situation similar to that reported in Mexico is feared, where native maize for human consumption in 33 communities of peasants and indigenous people (24% of the sample) from nine states was found to be contaminated with GMOs, mostly Bt. An analysis done on more than 2000 plants from different plots showed contamination ranging from 1.5% to 33.3% (CECCAM, 2003).

12.4 The Present Situation

Diverse environmental characteristics allow for organic production on a scale varying from sugarcane farms of a few hectares in north-eastern Argentina, to temperate zone farms of 1000 ha or more, to wool and lamb production on *estancias* of hundreds of thousands of hectares in the cold deserts of southern Patagonia (see Box 12.1). Because of extensive production in the low-yielding arid Patagonian steppes, Argentina is third in the world in certified organic area, but among the lowest in productivity: only US$11/ha. Fresh fruits, cereals and oilseeds are the principal export items; olive oil and wine specialities are increasing, with 90% of the production going abroad.

After a decade of rapid growth in production and exports, the boom is levelling at relatively low volumes because of a national economic crisis and increasing competition from new production areas in the temperate northern hemisphere. Moreover, processed organic food has not found acceptance among export customers, who as usual prefer to process the primary products in their own countries. Protectionist policies in the EU and the USA make the introduction of Argentinian products even more difficult. In 2005, Argentina's commercial organic

> **Box 12.1.** The extremes of the size range for organic farms in Argentina.
>
> **Sheep for Benetton**
> In the early 1990s, the Benetton Company bought two *estancias* for raising sheep: Cóndor (280,000 ha) and Coronel (335,000 ha), both in Santa Cruz Province in the far south of Argentine Patagonia, near the Strait of Magellan. This is a cold, dry environment. On average there is one sheep per 2.5 ha in Cóndor and one per 4 ha in Coronel. The vegetation is mostly low, perennial, cushion-shaped shrubs, and soil cover is scarce. Desert pavements are not uncommon. In just a century of extensive sheep grazing, the vegetation reached a stable level of degradation from which it does not seem to be recovering, not even in reserves. In the rainier west, along the Andes, a pine plantation is slowly being established (500 ha/year) to control erosion on the slopes.
>
> The owners decided to certify both *estancias* as organic (they have conventional ones in other Patagonian provinces, amounting to 900,000 ha), because they found it feasible to achieve sustainable grazing. They do this by means of rotations and periodic estimates of the grazing capacity of each parcel to prevent overgrazing.
>
> The company has its own genetic development programme based on Merinos and Corriedales. However, they do not certify the wool they use in their garments because there is no demand for organic wool.
>
> **Very small cane growers**
> In tropical Northern Misiones Province, 600 small tenant farmers grow 1500 ha of organic sugarcane, which they process in their own mill. They are descendants of central European immigrants who arrived a century ago.
>
> When the conventional privately owned mill that processed their cane went bankrupt in 1996, the provincial Institute for Farm and Industry Promotion, with the financial support of the national Secretary of Agriculture, sponsored the formation of a cooperative. The small tenants were trained in organic production and certification procedures. Most cultivate up to 1 ha of cane.
>
> They managed to lower gross production costs by 13% compared with conventional sugar producers. They diverted part of those costs to labour, increasing the labour input by 54% and thereby providing a better standard of living for the families joining the cooperative. The product is shipped mostly to Europe, with part going for local use. At present they are involved in a programme of farm diversification sponsored by two Italian NGOs: the Italian Institute for Foreign Cooperation and the Association for Rural Cooperation in Africa and Latin America.

production was only about 0.16% of the world total, and the future for most crops is unpromising, unless local markets reach full development and government policies specifically become involved. Only large traders, already having found their place in foreign markets, have been successful in adapting their production and exports, mostly with some fruits.

Organic activities have not found much support from the authorities, but export taxes have been lowered for organic products. The Argentinian government intends to back technological developments to process food and to sponsor some applied research in field technologies and plant health, so as to be in a better position to compete. A programme – not yet under way – that is focused on these aims has been proposed by MAPO, to be coordinated by the Inter-American Institute for Agricultural Cooperation and funded by the World Bank through the Programme for Provincial Agrarian Services. Also, INTA is considering setting up a programme of organic research and development for commercial farmers, and together with MAPO has been backing the development of groups of farmers with common crops and interests. Hopefully, these efforts will not come too late to aid Argentina's organic development.

12.5 Some Lessons from History

By contrasting the Argentinian situation to that of other Latin American countries where organic production of some crops has been successful – which has not happened very often – the following points stand out:

- Unlike in countries such as Chile or Brazil, after an initial period of official promotion of organic activity in Argentina, support has been dwindling. Official organic research and development have been nearly non-existent until now. Strong official support is needed from the start to get organic activity under way, including effective, steady promotional campaigns – which generally are expensive – in local and foreign markets. This is simply too great an effort for small organic NGOs.
- For small and medium-sized organic tenants, the only way to gain markets is through farmers' associations, a practice that is growing too slowly in Argentina's culture.
- Concerning international markets, given the long distances between producers and consumers, and with the prices of many organic crops dropping almost to commodity levels, only some market niches seem promising, namely, crops that do not compete with production in temperate countries or that are produced during the northern hemisphere's winter.

References

CECCAM (2003) Contaminación transgénica del maíz en México: mucho más grave. Press release, México, 9 October 2003. Centro de Estudios para el Cambio en Campo Mexicano; Centro Nacional de Apoyo a

Misiones Indígenas; Grupo de Acción sobre Erosión, Tecnología y Concentración; Centro de Análisis Social, Información y Formación Popular; Unión de Organización de la Sierra Juárez de Oaxaca; and Asociación Jaliscience de Apoyo a Grupos Indígenas.

Eco-Agro (1992) *Agricultura Orgánica, Experiencias de Cultivo Ecológico en Argentina*. Edit. Planeta Tierra, Buenos Aires.

Foguelman, D. and Lockeretz, W. (eds) (1999) *Organic Agriculture: The Credible Solution for the XXIst Century*. Proceedings of the 12th International IFOAM Scientific Conference. IFOAM, Tholey-Theley, Germany.

Foguelman, D. and Montenegro, L. (1999) Organic production and farmers in Argentina. In: Foguelman, D. and Lockeretz, W. (eds) *Organic Agriculture: The Credible Solution for the XXIst Century*. Proceedings of the 12th International IFOAM Scientific Conference. IFOAM, Tholey-Theley, Germany, pp. 45–50.

INTA (1992–2004) *Cartillas de PROHUERTA*. Buenos Aires, various dates.

Pais, M. (compiler) (2002) *La Producción Orgánica en la Argentina*. Edic. MAPO, Buenos Aires.

13 NASAA and Organic Agriculture in Australia

E. Wynen[1] and S. Fritz[2]

[1]*Principal, Eco Landuse Systems, 3 Ramage Place, 2615 Flynn, Canberra, Australia;* [2]*Catchment Officer – Biodiversity. Southern Rivers Catchment Management Authority, 1194 Wattamolla Rd., Berry, New South Wales 2535, Australia*

13.1 Introduction

In a country where organic agriculture is highly fragmented and its practitioners isolated, sooner or later there will be a movement to try to combine the related forces and present a united front towards the outside world. In Australia, this happened in the early 1980s, when the National Association for Sustainable Agriculture, Australia (NASAA) played a central role in the development of the organic industry in Australia. It set itself the tasks of defining organic agriculture; of providing information about the relevant issues to producers, consumers and the public; of being an intermediary in the market by providing certification and information to sellers and buyers; and of lobbying the government to provide a climate in which it was easier for organic agriculture to thrive. Here, we explore the history of NASAA and the significant contributions it made to the development of the organic industry in Australia.

13.2 Background

13.2.1 Agriculture in Australia

Australian agriculture is extensive in nature and export-oriented. These features also influenced the development of the organic sector.

The authors thank Stephanie Goldfinch, Ruth Lovisoli and Tim Marshall for their comments on an earlier draft of this chapter.

©CAB International 2007. *Organic Farming: an International History* (Lockeretz)

Of the approximately 15 million inhabitants in the early 1980s, around 80% lived in a handful of big cities, the capitals of each state. Large tracts of rural areas were, and still are, sparsely populated, with many areas being farmed extensively. This reflects the absence of adequate rainfall or irrigation.

In 2003/04 the total farm area, including pastoral properties but excluding horticulture, was approximately 443 million ha, with 86,700 producers. More than half of these ran livestock only (graziers). The second largest group, one-third of all producers, consisted of grain growers and grain-livestock farmers, and about 11,000 or 12.8% were dairy farmers (Australian Bureau of Agricultural and Resource Economics (ABARE), 2006, Tables P1–P8). In 2002, a total of 22,500 growers managed horticultural enterprises. Of the total agricultural production in 2005/06 of AUS$38.4 billion, almost 80% (AUS$30.4 billion) was exported (estimates by ABARE, 2006, Tables 3 and 6).

13.2.2 The growth of organic agriculture since the 1980s

Organic farming has grown rapidly from its small beginnings in the early 1980s. The number of organic farmers increased from fewer than 500 in 1982 to between 950 and 1200 in 1990, although the numbers are not strictly comparable (Conacher and Conacher, 1991). Another study (Hassall and Associates, 1990) reported that half of all Australian organic producers (600–700) in 1990 said that they had been farming organically for 3 years or less, with over two-thirds farming organically for 5 years or less. In 2005, the number of organic producers had risen to 1869, and estimates of the area under organic management put the figure at 11.8 million ha (Ian Lyall, AQIS, November 2006, personal communication) (the actual value is somewhat lower, as some producers are doubly or triply certified). This is large in absolute terms, but the fraction of total agricultural area under organic management is 2.6%, which is in the middle of the range for industrialized countries around the world. Figures provided by the two largest organic certifiers in Australia (NASAA and the Biological Farmers of Australia, BFA) indicate that approximately 97% of the total certified area was under extensive grazing management in 2005. This means that of the total of 11.8 million ha, close to 370,000 ha are in non-pastoral areas, which is approximately 0.7% of the total conventional area for included industries (the total area of wheat and other crops, mixed broadacre, and dairy for 2003/04 was 60 million ha, which does not include horticulture). Although the non-pastoral portion is only 3% of the total certified area, more than half the total value of the organic sector originates from those areas.

The growth of the market has paralleled the increase in the number of organic farmers. Fritz (1991) estimated the market for organic produce in Australia in 1987 at AUS$6 million. In 1990, the retail organic market was estimated at AUS$39 million (Hassall and Associates, 1990). The current retail value is a matter of dispute. Wynen (2003), using data from the certifying organizations, estimated the total farmgate value of organic production in 2000/01 at AUS$89 million, including produce not sold as organic. This figure was then used to estimate the retail value of that production. The result, reduced by the part sold in the conventional market, was AUS$107 million. To arrive at the total domestic retail value, exports should then be deducted, imports added and value added for processed goods. No details were available to do this, but it seemed reasonable to assume that exports from Australia were considerably higher than imports, and that processing was not a substantial part of the organic market. If this is correct, the retail value would have been closer to AUS$100 million than to the AUS$400 million at which NASAA put the retail value in 2003 (NASAA, 2003). For that year, Halpin (2004) estimated the total retail value of products sold on the organic market at AUS$127.9 million (estimated by adding all enterprises reported by the producer respondents). Imports were estimated at AUS$13 million, while export figures were (and still are) only available for quantities, not values (Halpin and Sahota, 2004). The NASAA figure may be high, because it presumably assumed continued growth at the rate of approximately 25% found between 1990 and 1995 (Hassall and Associates, 1995).

Exports have been an important part of the Australian market. Austrade (2003) estimated that one-third of Australian organic products were exported, with an export value of around AUS$50 million (year not specified).

13.3 The National Association for Sustainable Agriculture, Australia

13.3.1 Origin

In the early 1980s, no organic certification scheme existed in Australia. Organic agriculture was of interest to two different groups. The first consisted of farmers who used practices generally consistent with today's standards of organic agriculture. Most of these farmers were geographically isolated and did not know of the existence of other organic farmers (Wynen, 1990, 1994). Many had experienced significant problems with their own health or that of their crops or livestock when farming conventionally, and felt that drastic changes were needed to solve those problems. Later, when organic farming became better known, they found themselves fitting into a recognized agricultural sector.

Biodynamic farming was organized under the leadership of Bob Williams and Alex de Podolinski well before the 1980s. Later, some biodynamic farmers were united under the leadership of Alex de Podolinski, who ran the Bio-Dynamic Research Institute (BDRI).

The second group consisted of regional and state-based organic gardening organizations such as Henry Doubleday Research Association in New South Wales and the Soil Association in South Australia. Because of the large distances between them, these organizations usually operated in isolation. The public was largely unaware of organic and biodynamic farming per se.

Against this background, there was a perceived need for cooperation and for combining the efforts of all forces in organic agriculture. The idea of NASAA as a way to achieve this was first developed as a major project for a Graduate Diploma in Agriculture at Hawkesbury College, near Sydney (Fritz, 1984a). In 1983, Sandy Fritz circulated a proposal to develop a national association of organic agriculture to 13 organic organizations around the country, most of whom responded positively. Several articles were placed in organic and permaculture journals (e.g Fritz, 1984b). The idea of an umbrella organization that combined all forces interested in organic agriculture, including producers, consumers, traders and researchers, was presented by Fritz at several events in 1984, including the Organic 1984 Festival in Tasmania, the Permaculture International Festival in New South Wales and the Conference on Organic Farming in South Australia. The response at these events was generally enthusiastic.

13.2.2 Aims

As a result, a small group of people representing organic groups in all states of Australia began meeting to discuss the purpose and structure of a national association, and subsequently to develop a constitution. The general aims of such an association were to:

- Establish a communication network to assist organic growers in resolving common problems;
- Influence the direction of agricultural research and policy;
- Lobby to reduce policy and marketing obstacles to organic practices;
- Bring organic farming to the attention of the mainstream agricultural industry;
- Increase public awareness about organic farming.

Although many of the objectives were producer-oriented, it was recognized at a very early stage that organic agriculture could progress only if all stakeholders were involved, including consumers.

13.2.3 Structure

By early 1986, there was agreement on a constitution and a structure for the national organization, and NASAA was formally inaugurated. It was incorporated in early 1987, with Tim Marshall as its first chair and Sandy Fritz as secretary and de facto executive director. By that time, about 30 organizations, with a total of 5600 members, were affiliated with NASAA.

In those early years, the NASAA Committee consisted of two representatives from each state. These were chosen from State Councils, which were made up of two representatives from organic organizations in each state. Although in theory this was a democratic way to involve the grass roots, in practice it meant that those involved with NASAA had to be committed on several levels, and had to be able to travel long distances for meetings – a rather costly business in Australia. The obligation to be involved on several levels and the financial demands placed on individuals, such as for attending meetings, would prove too onerous for many.

This structure was ideologically based. Despite the recognized value of a broadly representative structure (such as being democratic and egalitarian, and promoting community networking), it also had important limitations. The structure and demands on the organization resulted in frequent turnover of directors. Additionally, the financial stress that any growing organization experiences necessitated much work to be done by volunteers, albeit professionals. Although NASAA's influence on the national scene increased in the late 1980s and early 1990s, the structure and geographic issues made it more difficult to achieve all that NASAA set out to do. Still, it enjoyed relative success in achieving its objectives.

By 1991, there was a recognized need to involve certified growers in decision making. The structure of NASAA was changed at that time so that certified growers elected one of the two representatives from each state. More recently (NASAA, 2000), it was decided that state representation was too cumbersome for the organization, and a new type of membership (voting membership) in NASAA was required in order to vote for Board directors. In 2006, NASAA had 126 members. This low level of community input, where not even all operators licensed by NASAA (such as growers and processors) have chosen to be voting members, perhaps indicates little interest in participating in decision making at the national level.

13.3 NASAA's Early Work

13.3.1 Setting standards

The first requirement for an organization that wished to inform the farming community and the public about organic agriculture and to influence government was to define organic agriculture. To this end, organic standards for Australian conditions had been worked on by a small group of people under the leadership of NASAA Committee member Lionel Pollard, who also developed Australia's Willing Workers on Organic Farms. The work on standards was brought to the NASAA Committee for completion. The first set of NASAA standards was in March 1987. NASAA then developed policies on certification (Wynen, 1989b), and a certification system was introduced, both for individuals and groups of producers, as well as for the service sectors, such as farm inputs and processing. At that time NASAA began to run workshops to train inspectors, and developed an inspector's guidebook (Marshall, 1990). It also supported work on conversion to organic agriculture (Wynen, 1992) and worked to promote the NASAA logo to ensure that consumers understood the importance of certification and to enhance consumer confidence in product integrity.

In the early 1990s, NASAA started to develop certification procedures for countries in Asia and the Pacific, especially at the request of European importers who used NASAA to do certification for them. In later years, this work expanded considerably.

13.3.2 Raising community awareness

NASAA's second requirement in those early years was to present the case for the importance of organic agriculture and to make this comprehensible for lay people, as well as policy makers. This was done by way of a book that constituted the first publication in a series of discussion papers (Wynen and Fritz, 1987). It outlined the problems with current farming practices, the alternatives and the policy implications of a shift to those alternatives. The book introduced many readers to NASAA and the scope of organic agriculture at the international level. It was launched by a federal politician and attracted significant media coverage.

At that stage, the organic movement used the word 'sustainable', as it was thought that 'organic' had negative connotations for many people outside the movement. In later years, the word 'sustainable' was appropriated by advocates of certain technologies within conventional agriculture, notably minimum tillage (also known as 'conservation farming', of which the main characteristic was the replacement of soil tillage with herbicides). The organic movement then went back to using the word 'organic', although NASAA has kept 'sustainable' in its name.

During those first years, a priority was to obtain media coverage not only of organic agricultural practices and events, but also of NASAA's stance on many issues of concern relating to conventional agriculture, including practices related to the use of pesticides and fertilizers, health issues and environmental impacts. In the late 1980s, NASAA started to be widely known in Australia, and was often sought out by the media for comments on these issues.

13.3.3 Education

NASAA's drive to provide information was directed not only at the media, but also at educational institutions. Before 1987, no agricultural colleges or universities in Australia were involved in education on organic agriculture in an official way. However, by 1991 several institutions had held conferences and established programmes on the subject.

NASAA seminars on organic farming were co-sponsored by at least five colleges of agriculture/horticulture. NASAA sometimes published the proceedings (e.g. Wynen, 1987) or provided funds. Hawkesbury Agricultural College established a 1 ha field dedicated to teaching organic production techniques. The Orange Agricultural College, now part of the University of Sydney, developed a postgraduate course in sustainable agriculture, with significant NASAA collaboration. This development included study modules in organic (Fritz, 1992) and biodynamic agriculture, a first for an Australian university.

The increased demand for knowledge on organic agriculture encouraged Australia's Technical and Further Education institutions to offer short courses on organic farming, sponsor guest lectures and hold seminars to educate their agricultural staffs. NASAA's first discussion paper (Wynen and Fritz, 1987) was an important source of knowledge, particularly for students looking for information on organics.

13.3.4 Marketing

In 1986, there was very little organic food available on the Australian domestic market. While some products were sold as organic, they were too few to constitute a real market. No significant wholesalers of organic products existed. However, by the end of 1987, several wholesale businesses specializing in organic produce were established in capital cities.

As a result of a stronger market, growers wanting to sell their products as organic began to seek certification to assure traders and consumers that their products truly were organically grown. Towards the end of 1989, applications for NASAA certification began to increase dramatically.

Around this time NASAA received a government grant (Fritz, 1987) to carry out a 12-month project to develop the market for organic products. The project focused on three areas:

- Developing the domestic market by increasing consumer awareness and information;
- Investigating processing requirements and opportunities for organic products;
- Investigating and developing export market requirements and opportunities.

Although NASAA was very clear that as an organic certifier it should not be involved with marketing of organic produce per se, it was also clear that a very important part of developing the industry was through helping organic farmers to develop the export market. An overseas marketing officer (Els Wynen) was appointed to provide information about NASAA and its standards and certification scheme to the European organic world so that importers would have confidence in Australian products with NASAA certification. The marketing officer also provided general market advice and export guidelines for Australian farmers who wanted to export organic products (Fritz and Wynen, 1991; Wynen, 1991).

Because exports are so important in Australian agriculture, government support for organic agriculture emphasized export promotion. For example, in 1989 the government funded consultants to research the organic market. NASAA assisted extensively in this work.

13.3.5 IFOAM accreditation

As assurance of organic status was essential to potential overseas buyers of NASAA-certified products, NASAA sought accreditation of its status as a certifying body by the International Federation of Organic Agriculture Movements (IFOAM). NASAA was evaluated by IFOAM in 1990, receiving a 'positive' evaluation; at that time, IFOAM did not yet 'accredit' organizations. Export markets responded favourably.

The IFOAM began offering full accreditation in 1993. NASAA was one of the first three certifying organizations to be accredited by IFOAM in 1994 (NASAA, 1995), along with KRAV (Sweden) and Biodinamico (Brazil).

13.3.6 Policy influence

NASAA's first discussion paper (Wynen and Fritz, 1987) was used extensively in the preparation of a Government White Paper (Parliamentary discussion paper) on organic agriculture by the Australian Quarantine Inspection Service (AQIS, 1988), part of the Federal Department of

Primary Industries and Energy, which was responsible for imports and exports. The White Paper included recommendations to immediately conduct a multidisciplinary study of the economic, social and environmental consequences of organic systems compared with conventional systems, as well as to develop a wide range of research and extension services for organic farming. Although such a study was never initiated, over time some work fitting this description was carried out with some government assistance, such as NASAA's market development work (1989), market research by a consultant (1990) and encouragement for state departments of agriculture to adopt a higher profile on organic agriculture.

The second NASAA discussion paper (Wynen, 1989a) detailed the marketing problems of organic wheat growers. The problems were caused by marketing regulations that compelled all wheat to be marketed via the Australian Wheat Board (AWB). As this organization did not have any provisions for handling and selling organic wheat, it sometimes entered into complicated arrangements with individual organic farmers, allowing them to market their own wheat. However, to take advantage of this possibility, wheat farmers had to meet the statutory requirements of the AWB, which in practice translated into payments to the AWB of up to AUS$30 per tonne (which could be between 10% and 20% of conventional wheat prices). At the time of the publication, a major overhaul of the entire wheat marketing regulation was taking place. As the existing policy was thought to discourage farmers from converting to organic management, NASAA published the discussion paper and lobbied the government. The new regulation allowed all farmers to sell their own wheat on the domestic market without a permit, and exports were allowed by other than the AWB, with permission. For organic farmers, this meant that they could now sell their wheat without the difficulties experienced in previous years.

The third NASAA discussion paper (Wynen, 1989b) concerned internal policies in setting certification costs. This formed the basis of NASAA's costing policies, and was published to promote transparency within the organic sector.

NASAA also provided submissions to, and appeared before, several commissions and enquiries into agricultural issues, such as the Royal Commission on Grain Storage, Handling and Transport (1988); Enquiry by the Industries Assistance Commission into the wheat industry (1988); and the Senate Select Committee on Agricultural and Veterinary Chemicals in Australia (1989). These submissions specified how existing policies affected organic farming, what changes would be needed to address the problems, and how organic farming could help solve the problems caused by conventional agriculture.

13.4 Developments from the Early 1990s

In the early 1990s, a major market for the Australian organic produce was Europe. The introduction of EC regulations in 1991 altered requirements for imports of organic products. This brought with it a need for official certificates to accompany imports into the EU. To meet this requirement, government accreditation of organic certification organizations became necessary. Hence, the Australian government became more important to the organic industry. The increased importance of certification, and changes in NASAA's management in 1992, meant not only that NASAA changed direction (focusing on certification), but also that other organizations emerged as providers of organic certification. In the resulting absence of a single organic industry voice, several attempts were made in the 1990s to unify the industry, an issue that only recently seems to have come to fruition.

13.4.1 National standards and certification

In 1990, AQIS called a meeting of organic stakeholders to discuss, among other issues, the need to develop a national organic standard that would facilitate exports to the important European market. The group consisted of representatives from the organic certifying organizations, of which there were three by that time, NASAA, BFA and the BDRI. Also present were representatives of the Australian Commonwealth Minister of Agriculture; the Federal Department of Primary Industries and Energy; three state Departments of Agriculture in their capacity as representatives of the Standing Committee on Agriculture/Australian Agricultural Council; the National Farmers Federation (a lobby group of conventional farmers); representatives for the consumers sector (the Federal Bureau of Consumers Affairs, the Australian Federation of Consumers Organizations and the Australian Consumers Association); and the Organic Retailers and Growers Association of Australia. At that meeting it was agreed to continue cooperation by formalizing the meeting's participants as the Organic Produce Advisory Council (OPAC). OPAC was to draft minimum national standards and inspection guidelines, and advise the Minister of Agriculture on matters of organic farming. OPAC agreed to minimum national standards in late 1991, which were endorsed by the minister.

These standards were referenced in a Ministerial Export Order to give them the force of law from 1 January 1992. However, they applied only to exports, not to the domestic market. In other words, within Australia the word 'organic' was not legally defined. As a result, Australian products that were not certified organic could be sold on the domestic market as 'organic' without legal repercussions. Consequently, because of World Trade Organization rules relating to national treatment, the Australian government could not require organic imports to be certified to any particular standard. Although the export standards served as the de facto domestic standards, uncertified products could be sold as organic in the domestic market. Despite several formal attempts to establish a legal definition of the term 'organic' for the domestic market, the issue was a matter of contention between the Australian organic industry and policy makers for a long time. Only recently (early 2007) has this situation changed (see Section 13.6).

The OPAC, as a representative body for the organic industry in Australia, was expanded to embrace all the certifying organizations accredited by AQIS for export purposes. In 2003, it was renamed the Organic Industry Export Consultative Committee (OIECC). Membership was changed to include government, the certifiers, the Organic Federation of Australia (OFA, see Section 13.6) and IFOAM. Membership was later extended to include the Organic Produce Programme of the Rural Industries Research and Development Corporation (RIRDC), a body that funds research on organic agriculture in Australia.

Contrary to the situation in most countries, the organic movement has had to pay 60% of the costs incurred by the government on its behalf. For example, for 2002/03 the AQIS programme, that is, the Australian national accreditation programme, cost the organic and biodynamic sectors AUS$84,500, to be paid by seven certifying bodies with a total of approximately 2345 total certifications, of which 1730 were certified growers. At present, it costs approximately AUS$105,000 per year, with a total of 2540 certifications, of which 1830 are certified growers (Ian Lyall, AQIS, November 2006, personal communication).

13.4.2 Certification: a changing role for NASAA

In 1992, considerable changes were made to the running of NASAA. Among others, some of the functions undertaken until that time by the secretary – the de facto executive director – were taken over by the chair. Some of the original committee members of NASAA, including Sandy Fritz, who held the position of secretary/executive director, left the organization.

Even though much of its work was still done by volunteers, NASAA experienced financial difficulties. Possibly partly to solve that problem, and because those with a bigger vision of the organization's role in the industry had left, NASAA focused on certification and moved away from the other tasks it had set itself, and which continue to be in its constitution.

Although the scope of NASAA's work has narrowed, it has done an excellent job in moving from a voluntary to a self-sufficient, non-profit organization, and has obtained a very good name internationally as a reliable organic certifier. Under the direction of Rod May and Jan Denham, NASAA has continued to stay well ahead of developments in the field. For example, in 2004 it became the first IFOAM-accredited certifier to achieve ISO 65 accreditation by the IOAS, an accreditation of the certification system as such that does not cover the standards. NASAA has been instrumental in consolidating contemporary practices within its own organization and also outside of NASAA, such as by establishing independent inspector services for the organic industry within Australia.

NASAA has had an important role in the international organic field, with representation on the Australian delegation to meetings on organic standards and certification of the Codex Alimentarius Committee on Food Labelling, which take place annually in Canada. NASAA representatives have also been active on IFOAM committees, such as the Programme Evaluation Committee and the Standards Committee.

13.5 Continued Unifying Efforts

One of NASAA's original main aims was to unify the organic industry. However, this proved to be a difficult task. Although the NASAA directors worked closely with a range of farmers (large and small), they themselves tended to be small farmers and academics in the first years of NASAA's existence. In the late 1980s, some large-scale organic farmers outside of NASAA decided to form a separate group representing farmers only. This resulted in the formation of the BFA, which adopted the NASAA standards with only minor changes. The BFA presented itself as an organization to promote organic farming by farmers, but from a different angle could be seen as 'early adopters' wanting to protect their interests, such as price premiums, by having more direct control over operations within the industry. The BDRI certified only biodynamic farms. At present, the two main certifiers (NASAA and the BFA) certify both organic and biodynamic producers.

From the early 1990s, several other organic certifying organizations emerged in Australia, including the Organic Vignerons Association of Australia (OVAA), which merged with the BFA in 2001; the Organic Herb Growers Association (which later changed to the Organic Growers Association (OGA), and presently is in the process of merging with the BFA); the Tasmanian Organic-Dynamic Producers (TOP); the Organic Food Chain (OFC), an offshoot of the BFA; and the Safe Food Production Queensland (SFPQ). The Organic Retailers and Growers Association of

Australia provide an industry-based certification programme for retailers and wholesalers.

Of the six remaining AQIS-accredited certifying organizations, four are listed under European and Swiss law, and as such can provide inspection and certification services for all Australian export consignments; five organizations provide inspection and certification services for products exported to Japan; two have 'conformity assessment' arrangements with the US Department of Agriculture's National Organic Programme, while other countries, such as New Zealand, Korea, Malaysia, Thailand, Singapore and Canada, currently accept Australian-certified produce that has been issued a government organic export certificate to verify its authenticity (Jenny Barnes, AQIS, November 2006, personal communication). At present, no foreign certification bodies are operating in Australia, and no local certification bodies work in association with international certification bodies.

With the increase in certifying organizations, the need for a unified voice for the industry was as important as ever. In 1992/93, the Department of Primary Industries and Energy funded a workshop to ascertain the priorities of the industry. The main recurring theme at the workshop was that of industry unity. At the end of the workshop, an Investigative Group was formed to assess options for one unifying structure, to develop proposals to assist future development, and to build on the results of the workshop (Wynen and Fritz, 1993). No single industry body was agreed upon at that time. In 1996, funding for a second workshop was provided by the RIRDC, and again, the call for unity was loud (Dumaresq et al., 1996). As a consequence, the OFA came into existence a year later, with the establishment of an interim committee in mid-1997.

13.6 Organic Federation of Australia

The interim committee comprised three certifier representatives, two growers, one processor, one wholesaler/exporter, one retailer, one consumer and an independent chair. The initial funding for the OFA came from the RIRDC, on condition that the R&D Committee of the OFA take the role of assessing and approving RIRDC grants for sustainable agriculture. The aims were similar to NASAA's original aims: unifying the industry by providing a forum for discussions; providing information; developing policies for organic agriculture; and lobbying the government. Issues pertaining to certification were left to the OPAC and the certifying organizations.

In its early stages, the OFA enjoyed considerable media attention, with its biannual scientific conferences (in 2001 and 2003) and its stand against genetically modified organisms (GMOs) as a major issue. None of the Australian organic standards allowed the use of GMOs, and the organic movement was lobbying conventional agriculture also not to accept their introduction. However, this topic was taken up less vigorously over the years, possibly in part because of a change to a chairman with different priorities, and in part as a result of bans on GMOs in many states in Australia, thereby creating the perception that the need for action was less urgent.

In 2004, a round table took place during which it was decided that the OFA's constitution needed to be amended to adapt the rules on representation by the different organizations in the OFA. By June 2005, the OFA adopted a new constitution. The structure consisted of a Main Board and several Advisory Boards, representing producers, consumers, certifiers, processors, traders, as well as the research and educational sectors. Organizations could join and send representatives to Advisory Boards, recently renamed Member Councils, which in turn had representatives on the Main Board. The aim was that the different stakeholders would decide what was important for them, and then get the weight of the OFA behind them to reach their aim. The OFA is represented on national committees.

One of the main issues for the revamped OFA was that of domestic organic standards. After lobbying the government long and hard for a change in this situation, the decision was made to house the organic standards in Standards Australia (a private, not-for-profit organization). This allows the government to call up these standards into regulation.

Another national concern is that of a national logo. This was an issue of special interest to consumers, traders and wholesalers, but less to certifiers. AQIS has offered the organic operators the use of a national logo, which is used by some, but by no means all.

These two issues, of domestic standards and a national logo, have really tested the notion of unity within the organic movement over the past few years. Although many in the industry profess that unity is important, in practice historical attitudes have prevailed whereby organizations seem to have a desire to retain their individual identity instead of forming a coalition that would advance common areas of national interest. In addition, in a world where powers shift, in this case away from the once all-important certifiers to the organic community in general (including consumers, marketing and education), it is perhaps not surprising that there are struggles to define boundaries. This factor inhibits the organic industry's ability to work effectively on the policy level with the government and within mainstream agricultural bodies, such as the National Farmers Federation.

Another major issue with which the OFA is involved is the direction of research in organic agriculture. Research funding specifically for organic agriculture provided by the RIRDC has only been around AUS$270,000 annually for several years (Wynen, 2003). The first attempt to get much greater funding was aimed at developing a Cooperative Research Centre (CRC) in 2002. A CRC is a consortium of different stakeholders, public and private (such as farmers, wholesalers, retailers and researchers) operating in a particular field, with considerable government funding commitments. In 2002, the efforts by proponents for an organic CRC were unsuccessful. Another attempt was mounted in 2004, where participation from a large retailer increased the likelihood of success. The OFA supported the proposal of the CRC. However, in 2004 the proposal was rejected, and little has come out of it since then. In 2006, the OFA published a position paper on its priorities on research and extension in organic agriculture in Australia (Wynen, 2006).

13.7 Concluding Observations

From the early 1980s, some people perceived that a single body could increase the organic industry's potential. Thus, NASAA was established. This organization had several priorities: defining organic agriculture (standards and certification); promoting organic agriculture via the media, educational institutions, conferences and seminars; communicating about how to convert to organic farming; seeking policy and research support; and facilitating marketing of organic products.

From the mid-1990s NASAA's change of emphasis and the establishment of other certifying groups resulted in an organization with a narrower scope of interest than was originally intended. In practice, emphasis shifted towards certification and away from education of the public and political lobbying.

The Australian government has shown little interest in organic farming, except regarding overseas market requirements. This is seen in the low level of research funding (via RIRDC) and the policy of having industry pay some of the expenses incurred by the government for services provided (e.g. by AQIS). The government's lack of interest has made involvement in issues other than certification difficult for private organizations, and certainly not financially sustainable.

Throughout the period of development, fragmentation has diminished the impact of the industry in gaining more supportive government policy, required agricultural research and commercial development. Repeated efforts to unite the industry occurred during the 1990s, ultimately resulting in the establishment of the OFA, indicating that many in the industry still see NASAA's vision of one industry body as important. It can perhaps be said that in a way, NASAA, through support of the OFA ever since its inception, has tried to realize some of its goals through this organization.

It is not clear why the organic industry's vision of unity has not been realized yet – whether the problem is the struggle for survival, the desire for success within individual organizations or personality conflicts among key players. What is clear is that the outcome is continuing competition among some organizations and their key individuals instead of cooperation. In recent work carried out on the adoption rate of organic agriculture in several European countries, cooperation or constructive competition among organizations serving the organic community was found to be an essential part of the industry's growth (Moschitz *et al.*, 2004). Organic industry bodies in Australia, although on their way, have yet to achieve this.

References

Austrade (2003) Organics overview – trade statistics. Available at: http://www.austrade.gov.au/australia/layout/0,0_S2-1_CLNTXID0019-2_-3_PWB1106308-4_tradestat-5_-6_-7_,00.html

Australian Bureau of Agricultural and Resource Economics (2006) Australian commodities. Available at: http://www.abareconomics.com/australiancommodities/commods/farmsector.html

AQIS (Australian Quarantine Inspection Service) (1988) The implications of increasing world demand for organically grown food. Government White Paper prepared for the Primary and Allied Industries Council, Canberra, Australia.

Conacher, A. and Conacher, J. (1991) An update on organic farming and the development of the organic industry in Australia. *Biological Agriculture and Horticulture* 8, 1–16.

Dumaresq, D., Greene, R. and Derrick, J. (eds) (1996) *Proceedings of a Symposium on Research Needs in Organic Agriculture*. Rural Industries Research and Development Corporation and Department of Geography, Australian National University, Canberra, Australia.

Fritz, S. (1984a) A national organisation for organic agriculture. Diploma thesis, Hawkesbury College, New South Wales, Australia.

Fritz, S. (1984b) Proposal 2. *Permaculture Journal* 15, 18–19.

Fritz, S. (1987) Marketing products of sustainable agriculture, an expanding niche in the marketplace. Submission on behalf of NASAA to Innovative Agricultural Marketing Committee, Export Development Group of Australian Trade Commission.

Fritz, S. (1991) Organic agriculture in Australia and the role of NASAA. *Ecology and Farming* 2, 25–27.

Fritz, S. (1992) *Organic Agriculture*. Sustainable Agriculture Course Module, UNE Orange Agricultural College, published by University of Sydney, Sydney, Australia.

Fritz, S. and Wynen, E. (1991) Guide to exporting products from Australia. NASAA, Adelaide, Australia.

Halpin, D. (2004) A farmlevel view of the Australian organic industry. In: *The Australian Organic Industry: A Profile*. Department of Agriculture, Fisheries and Forestry, Canberra, Australia.

Halpin, D. and Sahota, A. (2004) Australian food exports and imports. In: *The Australian Organic Industry: A Profile*,

Department of Agriculture, Fisheries and Forestry, Canberra, Australia.

Hassall and Associates (1990) Summary report on the market for Australian produced organic food. Report prepared for Australian Special Rural Research Council, Canberra, Australia.

Hassall and Associates (1995) The market of organic produce in Australia. Report prepared for the Rural Industries Research and Development Corporation, Canberra, Australia.

Marshall, T. (1990) *NASAA Inspectors Guidebook*. NASAA, Adelaide, Australia.

Moschitz, H., Stolze, M. and Michelsen, J. (2004) Report on the development of political institutions involved in policy elaborations in organic farming for selected European states. Report carried out with financial support from Commission of the European Community.

NASAA (1995) NASAA achieves international accreditation. *NASAA Bulletin* 2(4), 1.

NASAA (2000) The NASAA constitution updated! *NASAA Bulletin* 7(2), 1.

NASAA (2003) Australian research. *Technical Bulletin for NASAA Certified Operators* 1(4), 11.

Wynen, E. (ed.) (1987) *Sustainable Agriculture, A New Direction*. Proceedings of a symposium organised in cooperation with Dookie College, Victoria, Australia.

Wynen, E. (1989a) Wheat marketing and sustainable producers in Australia. Discussion Paper No. 2. NASAA, Sydney, Australia.

Wynen, E. (1989b) NASAA production standards implementation scheme: inspection, licensing and levies. Discussion Paper No. 3. NASAA, Sydney, Australia.

Wynen, E. (1990) Sustainable and conventional agriculture in southeastern Australia – a comparison. Economics Research Report No. 90.1. School of Economics and Commerce, La Trobe University, Bundoora.

Wynen, E. (1991) *NASAA Marketing Papers Series*. Information on exports of organic products to Northern America; Importers of organic products in Europe; Insights in the organic fruit and vegetable market in Europe; The organic grain market in some European countries; International trade of organic products with special reference to Australia.

Wynen, E. (1992) Conversion to organic agriculture in Australia: Problems and possibilities in the cereal-livestock industry. NASAA, Sydney, Australia. Available at: http://www.elspl.com.au/OrgAg/Pubs/Pub-A-FP/OA-FP-A4-Conversion1992.HTM

Wynen, E. (1994) Bio-dynamic and conventional dairy farming in Victoria: a financial comparison. Eco Landuse Systems. Available at: http://www.elspl.com.au/OrgAg/Pubs/Pub-A-FP/OA-FP-A8-Dairy1994.HTM

Wynen, E. (2003) Organic agriculture in Australia – levies and expenditures (No. 03/002). Rural Industries Research and Development Corporation, Canberra, Australia. Available at: http://www.rirdc.gov.au/reports/org/02-45.pdf

Wynen, E. (2006) Priorities for research and extension in organic agriculture in Australia. OPA Position Paper, May. http://www.ofa.org.au/papers/OFAResearchPolicyProposal-May2006-%20Final.doc

Wynen, E. and Fritz, S. (1987) Sustainable agriculture: a viable alternative. Discussion Paper No. 1, NASAA, Sydney, Australia.

Wynen, E. and Fritz, S. (1993) Organic industry project definition workshop. Department of Primary Industries and Energy, Canberra, Australia.

14 FiBL and Organic Research in Switzerland

U. Niggli

Director, Research Institute of Organic Agriculture (FiBL), Ackerstrasse, 5070 Frick, Switzerland

14.1 Introduction

The Research Institute of Organic Agriculture (known even in English by its German initials FiBL), was founded in 1973 and is now the world's largest research establishment for organic agriculture. Located in Frick, Switzerland, it also has an affiliate in Frankfurt, Germany (founded in 2001), and in Vienna, Austria (founded in 2004). Frick employs 120 people, with a project funding of about €10 million in 2005 (more details are available at: www.fibl.net/english/index.php).

Although founded primarily to conduct research, FiBL also gives high priority to turning knowledge into agricultural practice through advisory work, training courses and expert reports, with its activities disseminated through magazines, data sheets, reference books and the Internet. Oriented towards Switzerland and Western Europe at first, FiBL now also has numerous projects promoting the development of organic research services, as well as advisory and certification services in Eastern Europe, India, Latin America and Africa.

The description that follows places the history and development of FiBL in the broader context of the overall growth and development of organic farming in Switzerland, especially the role played by science in this controversial area. The story is made more complex by the importance in Switzerland of not one but two major sources of organic farming (described in Chapter 2, this volume): the work of Hans Müller, Maria Müller and Hans Peter Rusch, and the biodynamic movement, which originated with Rudolf Steiner in Germany. Although these two approaches overlapped, they also offered conflicting ideas about organic farming.

As will be seen, the attitude of Switzerland's agricultural establishment towards organic farming has gone from initial rejection, to tolerance, and finally to recognition and integration into the mainstream.

14.2 How the Pioneers Provoked Criticism from Science

Switzerland's first organic farm was started in 1920 by Mina Hofstetter (Vogt, 2000). She was strongly influenced by the German reform movement (*Lebensreform*) and had a strong affinity for the pioneers of 'natural' farming and gardening, the Germans Ewald Könemann and Julius Hensel (see Chapter 2, this volume). Hensel experimented with different rock powders as natural fertilizers to cure the negative effects of mineral sources of nutrients. As a consequence, rock powders were used in Switzerland until the late 1970s, and several research projects of FiBL focused on them as long-term fertilizers and organically acceptable pesticides that killed insects and fungi on leaves by drying them out or irritating them mechanically.

Shortly after the start made by Mina Hofstetter, the first Swiss biodynamic farm was established in 1932. Five years later, the Association for the Biodynamic Agricultural Method was founded at Rudolf Steiner's spiritual centre, the Goetheanum at Dornach, Switzerland. At the Goetheanum, scientific work on biodynamic agriculture was initiated by Rudolf Steiner in the late 1920s and carried out by Lili Kolisko and Ehrenfried Pfeiffer. The first reports on this work were published in 1931. Kolisko and Pfeiffer's gardens and modest laboratories in Dornach marked the birth of organic farming research in Switzerland (Kolisko, 1934–1936).

Many sources of organic farming were predominantly concerned with lifestyle issues, with a strong inclination towards a philosophical, religious, social and economic renewal of society (see Chapter 2, this volume). This nourished the agricultural establishment's perception that organic farming is merely religion, a prejudice that lasted very long and made it difficult for the federal authorities in Switzerland to recognize a need for any scientific work or even to implement a legal framework for regulating organic farming.

Hans Müller, a member of the Swiss Parliament and very successful lobbyist for opposition small-scale farmers, became interested in biodynamic agriculture starting in 1946. Very soon he started to develop organic farming further, separating it from its anthroposophic background, 'without mystic ado in a form which can be practised by farmers', as he wrote in 1954 (Moser, 1994). His wife, Marie Müller, was a very committed gardener and teacher, experiencing the new ideas of organic farming from her practical work. In 1950, Hans Müller met Hans Peter Rusch, the German microbiologist and medical doctor who provided a holistic scientific framework for the Müllers' future work with farmers. From 1953 on, Rusch analysed soils from organic farms with his specially developed 'Rusch Test', which used either selective lactose dextrose agar to grow lactic acid bacteria from soil solution, or other agars to grow coliform bacteria. This test became essential for the further development of organic farming; it gave, as Rusch wrote, security to those farmers who wanted to convert to organic farming without bigger risks. 'The "chemical" agriculture had its analytical tests and its laboratories – organic farming had nothing comparable' before he developed his test, wrote Rusch in retrospect in 1974 (Vogt, 2000). Hans Müller and Hans Peter Rusch continuously attacked the Swiss federal state research centres as being the driving forces of the chemicalization of agriculture, provoking a long-lasting and occasionally heated debate between the organic movement and the scientific community.

In 1948, the Federal Research Institute for Agronomy in Zurich had already started a 5-year plant nutrition field trial with stone meal, which showed that there was no positive effect on yields that could justify spending money for it. The organic movement retorted that with a reductionist one-factor trial in an unconverted 'dead' soil, of course no effect could be expected. Until around 1970, the battle between the federal state research centres and the organic movement of Hans Müller had been ardent on both sides. It culminated in the fortunately vain attempt to legally ban the term 'biological' for food, with the argument that it might lead to defrauding consumers.

The reasons for this long-term controversy were many. The state research centres, once made aware of the negative impacts of intensification on the environment and food quality, wanted to develop their own and in their view scientifically sound approach to cope with these problems. Integrated Pest Management (IPM) and later Integrated Plant Production (IPP) were seen as representing an ecologically optimized conventional approach that did not force both farmers and researchers to jump into a completely different system. After 1970, agroecological and agroenvironmental concepts were introduced into research. The scientific dialogue between the scientific community and proponents of organic farming, especially Hans Müller, remained difficult. In 1977, Ernst R. Keller, Director of the Institute of Plant Production at the Agriculture Department of ETH Zurich wrote that 'only very recently had it become possible to discuss things with exponents of the different organic lines'. At that time the first PhD work on organic farming was completed at ETH Zurich (Graf, 1977).

The fierce rejection of the organic pioneers by the scientific community was also partially the formers' doing. Some organic proponents were careless in dialogues with scientists. The Rusch test, for example, which was used to characterize the quality of organic soils, was already not the state of the art in the 1960s: 'From a soil biology point-of-view the classification of soils according to one single group of microorganisms must be criticized. That only this preferred group of lactic acid bacteria should be responsible for the fertility of soils does not even correspond to the usually holistic approach of organic farming' (Keller, 1977, translated by Urs Niggli). Also very blurred was Rusch's concept of the cycle of a 'living substance' in soil, plants, livestock and humans – driven by organic manuring and not working when chemical fertilizers were applied (Vogt, 2000). It was not really built on sound science.

It was in this context of continuously debating the relevance and questioning the scientific basis of organic farming that FiBL was founded (see Fig. 14.1).

Fig. 14.1. In 1997, Hartmut Vogtmann started as the first director of FiBL in a small room on this farm in Oberwil, Basel District. (Photo courtesy of FiBL. With permission.)

14.3 The Long Way Towards Institutionalized Organic Research

Heinrich Schalcher, member of the Swiss Parliament, asked the Swiss government in 1970 to create a new experiment station for organic agriculture or to convert one of the seven state centres to organic agriculture. Philippe Matile, Professor of Plant Physiology at the Botanical Institute of ETH Zurich shocked his peers at the university and the state research centres with several newspaper articles explaining the need for experimental research on organic agriculture. Michael Rist, lecturer for Livestock Buildings and Ethology at ETH Zurich, one of the early livestock pioneers in organic farming, joined in.

It was no surprise that the government saw no need for an organically oriented experimental institute. 'The control of crop pests with nonchemical means has already become a focus of the state research institutes', the government wrote in its answer to Schalcher's parliamentarian request on 2 September 1970. None the less, in reality, the seven state research centres remained negative and their leaders and senior scientists used their prestige to continue fighting against the organic movement. In 1974, and again in 1983, they defeated a Swiss regulation on organic farming and food. It was not until 13 years later, in 1996, that Switzerland put a law on organic food and farming into effect, one of the last in Europe.

As a result of the negative reaction of the government to the idea of an organic state research centre, a private trust was founded in 1973 to establish FiBL, which started projects in 1974. While biodynamic farmers showed a strong interest in research, Hans Müller's organic movement kept a clear distance from the research institute. In his opinion, organic farming was already invented and needed no further scientific input. He kept this view until his death in 1988, with the organic movement suffering from his repeated conjuring up of the recipes of Maria Müller and Hans Peter Rusch, both long dead. Consequently, the research programme of FiBL was biased towards biodynamic work. The stubborn attitude of Hans Müller provoked a split in the organic movement, and in 1975 Werner Scheidegger, the prominent proponent of the modernization of organic farming, finally sought a collaboration with FiBL.

It became FiBL's role to involve individual scientists from the state research centres and from the Agriculture Department of ETH in organic farming activities. This has helped to overcome – very slowly but steadily – the hostility between established agricultural science and the organic movement.

14.4 Research Priorities of FiBL over the Last Three Decades

When FiBL started its activities in 1974, food contamination issues such as high nitrate content in vegetables or pesticide residues in fruits and vegetables frightened the public and were debated in the press. As organic farming was not using nitrate fertilizers and pesticides, it was considered as a possible solution for these problems (Fig. 14.2). Consequently, the feasibility and the economic performance of organic farming were the first research questions asked by the political authorities and taken up by FiBL. To answer questions regarding agronomic feasibility, an experimental comparison, the 'DOK trial', was started in 1978 (Fig. 14.3), financed by the Federal Office for Agriculture (D stood for biodynamic, O for organic and K for conventional farming).

As the DOK trial was a field experiment with randomized plots on a single site, it was not suited to answer general questions on economic performance, labour inputs or marketing strategies of organic farming. A whole-farm comparison network was therefore established, collecting that kind of data and analysing them in comparison to data on standard practices. This work was planned and carried out jointly by a federal state research institute and the FiBL. As in many other countries, economists were the first to show interest in organic farming.

Fig. 14.2. FiBL's first research projects included this study of how the nitrate content of lettuce is affected by mineral versus organic fertilization. (Photo courtesy of FiBL. With permission.)

Fig. 14.3. The long-term DOK trial started by FiBL in 1978 in Therwil, near Basel. Still under way, the trial compares biodynamic, organic and conventional systems. (Photo courtesy of FiBL. With permission.)

Continuing the comparative way of thinking, a survey of the fertility of dairy cows in conventional and organic herds and a study of nitrate contents in organic and conventional glasshouse and field lettuce were launched.

In contrast to early research activities characterized by the question 'how good is organic in comparison to conventional farming?' raising funds for projects that helped to improve the organic system or to develop novel techniques for organic farmers was considerably more difficult. The first such work involved the composting of manure, the development of green manure cropping systems in maize and feeding trials with laying hens in order to overcome protein deficiency. Later, studies were started on phytoalexins in organically grown crops. Phytoalexins, such as resveratrol in vines, are antimicrobial, low-molecular-weight secondary metabolites produced by the plant after contact with microorganisms (bacteria and fungi) in order to activate its defence system.

As the biodynamic farmers were more open to what FiBL did, imaging or 'picture-forming' methods were studied for use in research on crop quality. It was already possible to show that organic and conventional products had different crystal patterns with copper chloride crystallization. However, what biodynamic pioneers claimed to be a difference in vitality of organic foods could not be demonstrated at that time, which remains true today.

In later years, there were more research activities that helped improve organic farming techniques. The institute became active in biological crop protection, variety testing with emphasis on horticultural crops, slurry aeration, different methods of physical weed control, as well as reduction of leakage and gaseous losses of nutrients during composting. In crop protection, slugs, aphids, apple moths, European corn borer, the cherry fruit fly, plum fruit fly, cabbage white and others became FiBL's favourites.

In the mid-1980s, innovative studies on landscape diversity, rural development and regional scenarios modelling the conversion of bigger regions were initiated. These studies heated up the debate on the future strategy of agricultural policy in Switzerland and led to models for the new agricultural policy in 1992, where state subsidies were decoupled from productivity and tied to environmental benefits (cross-compliance).

An important boost for organic farming came on 1 November 1986, near Basel, when a huge hall of Sandoz (which later merged with Ciba-Geigy into Novartis) that stored 1300 t of chemical compounds caught fire. The water used to extinguish the fire flushed pesticides and other compounds into the Rhine and killed all life over a stretch of 250 km. People in Switzerland felt threatened, and millions of people living along the Rhine from Basel to the delta in Rotterdam saw how life and ecosystems are endangered by pesticides. Organic farming was suddenly discussed with respect and the multinational pesticide companies Sandoz and Ciba-Geigy invested money to develop biocontrol agents such as different *Bacillus thuringiensis* strains against insects. This was not for long, though, as biocontrol compounds were difficult to deal with commercially and not effective enough.

From 1990 onwards, research activities at FiBL continued to grow. Many efforts went into optimizing the production of very difficult crops such as apples, cherries, berries, grapes, vegetables and potatoes. This work became necessary because leading supermarket chains started to sell organic products. The large-scale market channels, where no personal contact occurs between the producer and the consumer, demanded a better appearance and a higher technical quality of the products and continuous supplies. Organic labels gained a boost in the supermarkets: between 1992 and 2005 the number of organic farms grew from 800 to 7000, and organic foods' market share grew from 0.5% to 4.5% (Rudmann, 2005). The booming sales influenced the research agenda of FiBL heavily, and the whole food chain was subjected to micro- and macroeconomic analyses.

The more organic produce outgrew its niche, the more important it was to know more about consumers' attitudes towards organic food. Markets and consumers became a focus of research at FiBL, with themes like the elasticity of prices, what consumers expect of organic food, how their real and reported purchasing habits differ and what makes the organic supply chain so expensive.

The 1990s also brought a considerable shift in public awareness, from green topics such as environment and nature towards more personal issues such as human health and food risks. The situation of farm animals also became a major concern, and vegetarianism grew rapidly. Previously, animal welfare had been completely neglected by the organic movement. In Switzerland, several associations and non-governmental organizations (NGOs) that were not specifically organic were very active in the 1970s and 1980s, making animal welfare a subject of public debate. Hans Müller's organic movement never wasted a word on animal welfare. Livestock were the producers of organic manure and an animal product was organic because the animals were fed organically. Now, for once, the organic movement was not in the forefront of food trends. FiBL put animal welfare on the agenda, and within 10 years, from 1985 to 1995, the organic and animal welfare standards and labels were finally merged.

These actions brought appropriate husbandry systems into focus at FiBL, including free-range rearing of pigs, pasturing of laying hens and economically optimized cattle production on pastures. Very soon, emphasis was given to the medical side of livestock. By its own standards, organic health management should give priority to the choice of robust races, disease prevention and medication with alternative/complementary drugs and techniques. To close the gap between the theory of the standards and the reality of practice, prevention by herd management, complementary medicine such as homoeopathy and phytotherapy and classical biocontrol techniques became a new research priority at FiBL, with emphasis on udder diseases, diseases of young animals and endo- and ectoparasites.

In a long-term perspective, research is getting increasingly systematic at FiBL. A good example is the food chain approach, which is applied in many research projects. It integrates all steps from the input industry to production, processing, marketing and consumption, and scrutinizes organic food systems' economic, social and ecological progress. Another example is research in redesigning organic farms and organic production techniques by 'organic' minimum tillage, using landscape elements such as hedgerows, wildflower field margins and rotation fallow as functional diversity to stabilize pests and diseases. The most recent research projects address vegetable and apple growing systems that are completely self-sustaining and input-free.

14.5 FiBL's Broad Commitment

One of FiBL's strongest points has been not its excellence in science, but rather the combination of research with a mission to promote organic farming. Surveying, experimenting, analysing and commenting are what mainstream researchers usually do. They may claim objectivity, but it is never reached because even the best scientists bring some subjectivity to their work. FiBL's mission was to build up capacities, structures and communication in organic farming and to organize the community. As a consequence, the dissemination of all kinds of information and know-how were also part of its mission. To bridge the gap between researchers and farmers, most of FiBL's research has taken place on commercial organic farms. A growing number of commercial farms participating in experiments have become crystallization points for training and information exchange among farmers and scientists.

Yet the commitment involved more than dialogue with farmers. Two years after the International Federation of Organic Agriculture Movements (IFOAM) was founded in Versailles (5 November 1972), Hartmut Vogtmann, then FiBL's director, joined up with IFOAM. He organized the second IFOAM meeting in Seengen in 1976 and the first International IFOAM Scientific Conference in Sissach in 1977, and hosted the IFOAM secretariat from 1975 to 1980 at FiBL (see Chapter 9, this volume, for the history of IFOAM). Standards-setting became a top priority for many years at FiBL, both on the national level with umbrella standards for Switzerland and on the international level, first with the IFOAM Standards Committee and later with Codex Alimentarius of the Food and Agriculture Organization (FAO) and World Health Organization (WHO). Once Switzerland had basic standards for organic farming, an umbrella organization was initiated called Bio Suisse. Its nationally known label, a stylized green bud, was first FiBL's own logo but was then given to the new organization. The next step to build up was certification. For more than 20 years, FiBL worked to develop methods and procedures that allowed inspection and certification of farms and food processors. Such an approach of tracking (process-oriented) was novel in food science but was necessary because the traditional way of traceing (product-oriented) could not positively distinguish organic from conventional foods. Once finished, the knowledge of inspection and certification was transferred in a spin-off to a private company.

This understanding of its role brought intensive stakeholder involvement to FiBL and made its work highly relevant for practice. It was an exciting piece of pioneer work and brought FiBL national and international recognition.

14.6 Has Organic Farming Become Mainstream?

With an annual per capita consumption of more than US$100 worth of certified organic food in 2003 (Richter and Padel, 2005), the Swiss are the world champions. All sectors of Swiss society have integrated organic farming in a positive way, and organic products have become common among food retailers and processing companies. The state schemes for research and advising have adopted organic food and farming as their most innovative pet. Already half the research on organic systems is done outside of FiBL at state research centres. Organic farming perfectly matches the philosophy of multifunctionality developed by the government to cope with the challenges of globalization and the requirements of the World Trade Organization.

Has FiBL's mission been accomplished? Certainly not! Debates on coexistence of organic farming and genetically modified organisms (GMOs) show that policy makers and the GMO industry accept organic farming as a niche, but have other plans for mainstream agriculture. The reality of worldwide markets still promotes a highly industrialized agriculture. None the less, organic farming has a huge potential for creating an ecologically, economically and socially sustainable farming and food system. This is a big challenge for researchers, too, and it is certain that there will be further debates among scientists. It is equally certain that FiBL will continue to participate in them.

References

Graf, U.R. (1977) *Darstellung verschiedener biologischer Landbaumethoden und Abklärung des Einflusses kosmischer Konstellationen auf das Pflanzenwachstum.* PhD thesis No. 052, ETH Zürich, Switzerland.

Keller, E.R. (1977) Biologischer Landbau – Alternative oder Denkansatz? *Neue Zürcher Zeitung* No. 91, 20 April, and No. 97, 27 April.

Kolisko, L. (1934–1936) *Mitteilungen des Biologischen Institutes am Goetheanum* Nos. 1–5. Dornach, Switzerland.

Moser, P. (1994) *Der Stand der Bauern: bäuerliche Politik, Wirtschaft und Kultur gestern und heute.* Huber-Verlag, Frauenfeld, Switzerland.

Richter, T. and Padel, S. (2005) The European market for organic foods. In: Willer, H. and Yussefi, M. (eds) *The World of Organic Agriculture. Statistics and Emerging Trends 2005.* IFOAM/FiBL, Bonn, Germany/Frick, Switzerland, pp. 107–119.

Rudmann, Chr. (2005) *Biolandbau und konventionelle Landwirtschaft im Vergleich.* In: Rudmann, Chr. and Willer, H. (eds) *Jahrbuch Biolandbau Schweiz 2005.* FiBL, Frick, Switzerland, pp. 27–44.

Vogt, G. (2000) *Entstehung und Entwicklung des ökologischen Landbaus im deutschsprachigen Raum.* Ökologische Konzepte 99. Stiftung Ökologie & Landbau, Bad Dürkheim, Germany.

# The Organic Trade Association

15

K. DiMatteo[1] and G. Gershuny[2]

[1]*Senior Associate, Wolf & Associates, Inc., 90 George Lamb Road, Leyden, Massachusetts 01337, USA**; [2]*Gaia Services, 1417 Joe's Brook Road, St. Johnsbury, Vermont 05819, USA*

15.1 The Origins

The story of the Organic Trade Association (OTA) mirrors the continuing growth and blossoming of organic agriculture and the organic products industry.

The Organic Foods Production Association of North America (OFPANA), as OTA originally was called, was born in 1984 – a watershed year for the organic community. Until that time there were just a few organic certifiers, primarily small grass roots grower groups. A few state governments had also become involved in certification.

Organic food was associated primarily with fresh produce, although grains and beans were increasingly finding markets. The biggest state for organic production was California, which supplied produce to most of the country, and the seal of the California Certified Organic Farmers, one of the country's oldest certifiers, was the only one recognized by many consumers.

In 1984, the picture began to change. That year a farm input dealer and the president of a produce distribution company, both from the mid-Atlantic region, headed north to Vermont to meet with a locally known and trusted farm consultant. They were looking for northern growers to supply organic produce in summer, when lettuce and other cool season crops were not available from mid-Atlantic and southern producers. Moreover, they wanted the crops to be certified organic. The idea they were pitching was a new, producer-controlled national certification

---

* When this chapter was written, Katherine DiMatteo was Executive Director of the Organic Trade Association, Greenfield, Massachusetts, USA.

©CAB International 2007. *Organic Farming: an International History* (Lockeretz)

programme. That first year, all certification costs were covered by the produce distributor, and 23 growers signed in – about as many as the total number that had been certified by the Natural Organic Farmers Association, the region's leading certifier, since it began in 1977.

In spring that year, word came that the International Federation of Organic Agriculture Movements (IFOAM) was sponsoring a meeting in the USA to try to bring together organic activists and merchants to discuss common issues, and perhaps form a North American association. It seemed that national organizing was in the air.

As many as 18 people from around the USA and Canada attended the meeting, including manufacturers, academics, grass roots organizers and certifiers. Four participants (including Grace Gershuny) were working on the pilot certification programme. Others included representatives of organic processors and distributors, a long-time small farm advocate, a representative of an important federation of cooperatives and several people from the trade association of the natural foods industry. Almost all these people, whether non-profit activists or entrepreneurs, had long been personally committed to organic production methods for the sake of their own health and the health of the planet. They all recognized that the organic idea was on the verge of taking off, but needed a unified, consumer-trusted market identity to fulfil its potential.

The group quickly agreed that a continent-wide organization was needed. Already there were multiple state laws and differing definitions of organic, conflicting standards among the handful of certifiers, and frustration among processors in finding consistent, reliable sources of organic products. The objectives of this new organization were stated as:

- Presenting a common image of organic foods in the marketplace;
- Establishing certification guidelines for organic foods;
- Evaluating and endorsing certification programmes and processes;
- Setting standards of excellence for the industry.

The group, tentatively calling itself the Organic Foods Production Association of North America, held its next meeting that autumn, and decided to incorporate formally as a non-profit association. (After many years of members' complaints about the unwieldy and forgettable name of the organization, it was changed to the Organic Trade Association in 1994.) Momentum grew for creating a viable trade association that could advocate for the interests of its membership. Funds for conducting its business, therefore, had to come from the members, and a few of the entrepreneurial founders took the initiative in soliciting contributions.

The organization's ambition and accomplishments were immense compared with the resources available: mainly volunteer time and personal funds. In 1990, the Board of Directors hired its first paid staff member (Katherine DiMatteo), who until March 2006 was the OTA's executive director.

From its modest beginnings two decades ago, the association has grown substantially, now boasting a staff of 15, a membership of 1500 and an annual budget of US$1.8 million.

15.2 Developing the OFPANA Guidelines, and the Idea of Accreditation

The top priority of the new organization was to develop a set of unified guidelines for organic standards – a consistent national definition that everyone involved in organic production would support. This would then form the basis for a consistent message to consumers and the public about the meaning of 'organic' on a food label.

The growing problem of competing certification programmes with conflicting standards would be addressed by establishing an OFPANA endorsement programme. The consensus was that consumers should be encouraged to look for a certifier's seal as their assurance that the organic label on a product was legitimate. The intention was to set up a process for evaluating the various certification programmes and allowing those that passed to claim the legitimacy of OFPANA's endorsement. It was also believed that this would allow mutual recognition among certifiers. Farmers would thus no longer need to carry multiple certifications in order to sell to diverse processors who required their farmer suppliers to be certified by the same programme as they were certified.

It was decided that the document that would embody this goal should be referred to as guidelines, not standards, because they were intended to serve as an umbrella for more specific regionally based standards. As guidelines they should be general enough so that local, membership-controlled organizations would retain control over the regionally specific standards they adopted. Certifiers would be able to recognize each other's standards as equivalent, provided they fell within the OFPANA guidelines.

Creation of the guidelines was a key accomplishment for helping OFPANA to emerge as the voice for the organic trade. The guidelines also set an important precedent because they outlined the certification process that should be used to verify compliance with standards. Finally, the open, transparent and democratic process by which drafts were circulated and approved by the membership was a crucial factor in establishing broad consensus and credibility for this foundational document.

The guidelines were revised in 1988, with the critical endorsement programme still undeveloped. One controversial piece of the original version that was removed during this round was the guideline addressing fair labour standards. Although many may regard this as signalling the evisceration of the social conscience of the organic industry, the consensus at the time was that matters of equity and labour relations did not belong in organic standards – the organic label could not be used to redress every problem of the food system, and enforcement would present major obstacles.

15.3 The Debate Over Organic Definitions

Developing the OFPANA guidelines involved collecting and analysing certification standards and programme information from every known certifier in the USA and Canada. The process identified the principles and values that formed the common basis of the standards, as well as any contradictions among them. Most were quite similar, as a result of having borrowed liberally from each other. A key objective was to establish organic production methods as scientifically credible, while remaining consistent with consumer expectations.

A fundamental tension emerged between an orientation emphasizing production systems versus the desire of consumers for 'food you can trust'. This tension continues to exist, as seen in the arduous deliberations of the US Department of Agriculture's (USDA) National Organic Standards Board (NOSB) and the bureaucratic demands of accountability within the realm of organic certification.

The first round of guidelines was production oriented, rooted in the concept of 'agronomic responsibility' as the basis for organic standards. This gives primary importance to the effects of a given practice on the health of the soil and the farm organism, and holds that product quality is an inevitable outcome of soil and agroecosystem health. By this reasoning, the origin of a given material does not matter. Synthetic compounds might be more environmentally benign than natural ones, and in any case it is often difficult to define clearly what constitutes a 'natural' material. Although many consumers clearly believed that organic meant 'chemical free' or 'no synthetics', all existing standards allowed some synthetic materials to be used, and prohibited or restricted the use of some natural materials, such as raw manure. Supporters of 'agronomic responsibility' argued that the credibility of the organic label required us to educate consumers rather than perpetuate a false image.

An opposing view argued for 'origin of materials' as the basis for organic standards, based on the idea that organic certification is primarily a consumer guarantee system. The rationale was that consumers had come to believe that organic producers used no synthetic inputs, and consequently believed in organic products as cleaner, purer and safer than conventional ones. The flexibility and need for judgement inherent in the 'agronomic responsibility' approach was considered dangerous because it opened the system up to abuse, as opposed to providing clear, bright lines and allowing for greater consistency in decision making.

By a narrow margin, OFPANA's members favoured the 'origin of materials' as the basis for organic standards. The board decided to change the OFPANA guidelines to eliminate all synthetic materials from the 'accepted' category, and to establish criteria by which synthetics might be considered acceptable on a case-by-case basis. This approach was codified in the Organic Foods Production Act (OFPA) of 1990, which gives the primary responsibility for reviewing and approving materials to the NOSB. Although both approaches allow some synthetic materials to be used, the 'origin of materials' approach is more restrictive, and results in fewer materials being allowed.

Besides whether to allow certain synthetic materials, such as dormant oil for fruit trees and copper sulphate as a fungicide, the question also arose as to whether some naturally occurring materials should be prohibited. Most controversial of these is the fertilizer Chilean nitrate, or nitrate of soda. Long prohibited by IFOAM and zealously opposed by European organic farmers, it was favoured by several northern US organic growers as a way to boost available nitrogen in early spring in cold soils. The Chilean nitrate debate is still with us, an enduring symbol of the dilemma posed by the 'origin of materials' criterion.

Position papers on several critical issues for organic standards development were developed during OFPANA's first decade. Among the most important of these was a 1986 paper entitled 'Laboratory Testing and the Production and Marketing of Certified Organic Foods'. This was prompted by the threat posed by emerging labelling schemes based on testing foods for the presence of pesticide residues, with certification as 'residue free'. Two other important ones were issued in 1992. One, on accrediting organic certification agents, presented the opinion that accreditation of certification agents should be based on the International Organization for Standardization (ISO) Guideline 61, 'General Requirements for Assessment and Accreditation of Certification/Registration Bodies'. The other, on biotechnology in organic agriculture, argued that genetic engineering was not equivalent to traditional breeding techniques, and that products of genetic engineering should be considered synthetic; therefore, in general, they should be prohibited in organic production and subject to review according to the criteria for allowed synthetic materials.

15.4 Committee Activities and the Evolution of the Organization

Early in OFPANA's life, a technical committee was formed to advise the board on development of the guidelines and the criteria for acceptable organic practices and materials. The most well-developed aspects of the guidelines were those addressing soil fertility and crop management. Livestock and food processing guidelines were much more general, reflecting the slower development of these sectors. For example, the USDA had a long-standing prohibition on labelling meat as 'organic'. This prohibition, which severely inhibited early development of the organic meat and poultry sector, remained in effect almost until the final version of the national standards was published in 2000. Organic processing likewise lagged crop production. In the mid-1980s there were few organic food manufacturers, and processing mostly involved simple milling and grinding of grains, washing and packing of fruits and vegetables or manufacturing of products such as bread, chips and cereals. Few certifiers even had standards for organic processing or manufacturing or offered certification to processors, and processed organic products were rarely seen in mainstream retail outlets.

As the need to develop more detailed guidelines for livestock and processing became apparent, the Technical Committee began forming specialized subcommittees. In 1988, the Humane Society of the United States began working with OFPANA to help define standards for humane livestock treatment.

Later that year several organically inclined nutritionists and food scientists began examining the various ideas regarding what constitutes organic processing. The MPPL – Manufacturing, Processing, Packaging and Labelling – became the next subcommittee of the Technical Committee. After extensively reviewing research about various processing methods and their effects on the nutritive content of food, and debating temperature ranges and the definition of 'minimally processed', the MPPL committee concluded that organic processing standards should only deal with protecting the organic integrity of the raw farm product. Processing itself should not be considered an organic practice – the basis of the organic claim rests with farm practices, and processors and manufacturers are charged with preserving the value added to the product by virtue of its ingredients having been produced using organic methods. The materials and methods used in processing are allowed or prohibited according to the same standards as crop and livestock production materials and methods, with no restrictions placed on what kinds of foods might be made with organic ingredients. (For example, although hydrogenated oils were prohibited, this was not because they were considered bad for one's health, but rather because the solvents used to produce them do not meet the criteria for materials or methods allowed in organic production.) While this decision raised the possibility of the dreaded 'Organic Twinkie', it also avoided entangling the industry in a whole additional arena of regulation and controversy over acceptable processing methods.

As the specialized concerns and complexities multiplied, so did the subcommittees. Today, various specialized groups within OTA are developing and reviewing standards for specific aspects of organic production and processing, including distribution, retailing, personal care products and fibre processing.

Another extension of OFPANA's activities was the formation of the Ethical Practices Committee in 1987 as a forum for resolving internal disputes and concerns within the industry in a non-public setting. It included representatives of all sectors of the trade. A process for mediating disputes between OFPANA members was created and applied in several instances, and the committee also developed a code of ethics for inclusion in subsequent versions of the guidelines. Today the OTA's Code of Ethics stands apart from the guidelines; it is regularly reviewed and revised, and acceptance of it is a requirement of membership in the OTA.

Finally, early efforts to create international linkages continue today under the International Relations Committee. Among many early initiatives was one to establish OFPANA as the North American 'wing' of IFOAM. Although this was unsuccessful, OTA remains a proud and active member of IFOAM. While the effort to create an endorsement programme never succeeded, OTA members involved with IFOAM contributed to the development of IFOAM's accreditation programme. With increased international trade in organic products and difficult questions of harmonization of international standards, the International Relations Committee has maintained a strong presence in organic trade negotiations everywhere, including meetings with the USDA's Foreign Agricultural Service and the Codex Alimentarius Commission.

15.5 Promoting Organic

Promotion of the generic organic concept was a primary purpose of OFPANA from the start. Various marketing committees produced promotional materials such as press releases, bumper stickers, buttons to be worn at trade shows, as well as brochures and posters with organic definitions for retail stores.

A great leap for the organization in marketing organic products was the Organic Harvest Month, organized annually since 1992. It includes distribution of consumer information as part of a well-orchestrated media campaign. The addition of a professional media and public relations staff has kept the public promotion of organic foods a centrepiece of OTA's activities.

Another major promotional activity is OTA's All Things Organic Conference and Trade Show, held annually since 2000. The idea of starting an all-organic trade show was the result of the increasing importance of organics at the country's largest natural foods trade shows.

15.6 The Grass Roots versus the Suits: OFAC, OFA and the Conflicts Among Industry Sectors

A major area of tension within the budding trade association had to do with the division between grass roots organizers, who tended to be idealistic, and the business people, whose primary concern was for expanding the organic industry. The historical tension between primary producers and the middlemen who reap the biggest share of the consumer's food bill contributed to a feeling of distrust among small farm advocates towards the 'suits'. The idea that the trade association could be a forum for collective negotiations was counterbalanced by the perception among many grass roots people that the organization was dominated by business interests, and that supporting it would only play into their (clearly suspect) agenda. This perception became a self-fulfilling prophecy of sorts, although ironically many of those involved in manufacturing, distribution and retailing considered OFPANA to be too heavily oriented towards the farm constituency.

One way the association has tried to provide a big tent for diverse constituencies has been to encourage the formation of sector groups. Farmers and farm advocates were involved from the beginning, and the organization attached considerable importance to farmer representation in its direction. The Organic Farmers Associations Council (OFAC) was established as an autonomous constituency group within OFPANA in 1989, and included representatives of many of the farmer-directed organic organizations. The OFAC's pivotal role in the passage of the OFPA, in collaboration with a coalition of consumer and environmental organizations, was a major success story in the history of grass roots lobbying.

Once the OFPA was passed and the urgency of the organizing mission receded, participation in the OFAC dwindled. Eventually, the certification organizations within the OFAC split off to develop their own industry sector group, the Organic Certifiers Caucus (OCC), joined by state and for-profit certifiers who were not members of the OFAC.

The OCC gained momentum from the need to represent organic certifiers in the development of the new regulations for the USDA's programme to accredit certifiers, as mandated in the OFPA. Although many of the original grass roots members of the OFAC became part of the OCC, this body was not devoted to the interests of producers; also, some key OFAC member organizations did not run certification programmes. It became clear that the farmers and the certifiers did not have the same interests and concerns. In the process, a unified voice for the farm sector within OTA was lost. In 2002, the board agreed that OTA needed to turn its attention back to its relevance for organic farmers.

Finally, another sector that became motivated to develop its own voice consisted of the companies that produce and sell supplies for organic production. The Organic Suppliers Advisory Council (OSAC) was formed within OTA in 1994. The variety and inconsistency of lists of approved materials developed by the certification organizations was a barrier to this sector's success. Eventually the efforts of OSAC in conjunction with the Organic Certifiers Council led to the establishment of the Organic Materials Review Institute, which today provides services both to input suppliers seeking acceptance of their products for organic production, and to certifiers needing to know whether the brand name products their clients want to use comply with the USDA's National Organic Program (NOP). OSAC continues to work to clarify and unify state and federal organic labelling requirements on farm and garden products.

15.7 Creating an Industry: the Continuing Quest for Reciprocity

The never-ending quest for unity within the industry was seen most clearly in repeated attempts to develop reciprocity agreements among certification programmes. In 1989, in the first of what would be a series of such meetings, a professional facilitator was hired to help with an agenda that was known to involve some intense internecine rivalries. Despite productive discussions and repeated pledges of cooperation, reciprocity agreements quickly collapsed once the meetings were over.

Reciprocity was also a subject of intense discussion at a meeting sponsored by the USDA for certifiers in 1996 to inform them of progress towards developing the NOP's proposed rules on accreditation of certification agents. However, despite the incessant discussions held, agreements circulated, bridges built and fences mended, the organic certifiers remained mutually distrustful, creating cumbersome verification procedures to permit manufacturers to accept different organizations' certification of ingredients in their products. OFPANA's efforts to bring the certifiers together failed. It finally took USDA accreditation and the force of law to bring about the reciprocity that had eluded the community.

15.8 Federal Rules

Passage of the OFPA in 1990 was undoubtedly the biggest milestone for the industry during the first 20 years of OTA, although there are those who have regrets about involving the federal government so strongly in regulating the industry. The association's work in the time since the OFPA was passed has focused strongly on serving as a watchdog of the regulations as they developed, keeping its members informed of progress and organizing the industry's response to the USDA's proposals.

Moving from law to regulation took more work and time than anyone ever dreamed in the heady days following the OFPA's passage. Year after year, the publication of a proposed regulation was 'imminent'. When the first proposed rule was finally put out for public comment in December 1997, OTA analysed it and held public forums to gather member input for response. Many consider this work to have been of primary importance both in establishing OTA as the leading voice for the organic industry and in raising public awareness about the growth and importance of the organic trade.

A major effort to completely revise the old OFPANA guidelines (which had served as the template for the NOP's proposed rule) was mounted in response to the anger generated by the USDA's proposal, especially the 'Big Three': the proposal's failure to exclude the use of sewage sludge, genetically modified organisms and food irradiation. The idea was to present the USDA with a set of organic regulations that reflected the consensus of OTA's members and that would replace the widely rejected standards proposed by the USDA. Thus was created the American Organic Standards (AOS).

The unprecedented extent of negative public comment – some 280,000 responses, overwhelmingly critical – led the USDA to withdraw the first proposed rule and rewrite it, a process that took until December 2000, when final standards were published. OTA and its AOS played an important role in creating a more acceptable alternative to the first proposed rule, without discarding those aspects that were considered consistent with the industry's needs. To many, OTA's ongoing monitoring of government's regulatory and market development activities is the primary benefit it offers.

15.9 Looking Back to Look Ahead

As the industry grows, OTA's membership, staff and challenges will also continue to grow. Those who have been involved for many years have valuable insights to offer. Many speak of increased education about the benefits of organic farming as the most important task for the next phase, pointing to the new non-profit Organic Center for Education and Promotion (now the Organic Center) as an exciting new activity. As Bill Wolf, the initiator of OSAC, puts it, 'OTA is continuing to be a unified and ethical voice', adding that 'the one challenge is establishing more ties to farms – getting back to earthworms and soil'. Fred Kirschenmann, prominent among early OFPANA figures, agrees that OTA must 'stay ahead of the curve – help organic fulfill its potential' by looking towards more local food system choices as the next wave of market development. Bob Anderson, an organic food industry pioneer and a founding member of the Organic Food Alliance (later incorporated into OTA), emphasizes that 'OTA must continue to be open to everyone – it's extremely important to get everyone to the table, no matter how divisive the issues are'.

16 A Look Towards the Future

B. Geier,[1] I. Källander,[2] N. Lampkin,[3] S. Padel,[3] M. Sligh,[4] U. Niggli,[5] G. Vogt[6] and W. Lockeretz[7]

[1]*Colabora, Alefeld 21, 53804 Much, Germany;* [2]*Ecological Farmers Association, Gäverstad Gård, 61494 Söderköping, Sweden;* [3]*Organic Research Group, Institute of Rural Sciences, University of Wales, Aberystwyth SY23 3AL, UK;* [4]*RAFI-USA, PO Box 640, Pittsboro, North Carolina 27312, USA;* [5]*Research Institute of Organic Agriculture (FiBL), Ackerstrasse, 5070 Frick, Switzerland;* [6]*Friedrich-Naumann-Str. 91, 76187 Karlsruhe, Germany;* [7]*Friedman School of Nutrition Science and Policy, Tufts University, 150 Harrison Ave., Boston, Massachusetts 02111, USA*

The agriculture of ancient Rome failed because it was unable to maintain the soil in a fertile condition. The farmers of the West are repeating the mistakes made in Imperial Rome. . . . How long will the supremacy of the West endure? The answer depends on the wisdom and courage of the population in dealing with the things that matter. Can mankind regulate its affairs so that its chief possession – the fertility of the soil – is preserved? On the answer to this question the future of civilization depends.

(Howard, 1940, p. 20)

The study of the history of organic farming reveals both continuity and change since the beginning. The continuity will allow us to define organic farming's core principles; the changes show how organic farming adapted to different ecological, social and political challenges. Reflecting upon organic farming's origins and locating the present-day situation within its historical context may increase our awareness of current problems and possible solutions.

These core principles of organic farming have not changed:

- Organic farming systems are based on a biological understanding of soil fertility that emphasizes the interactions among soil humus, soil organisms and plant roots, as well as on an ecological understanding linking plants, animals and humans.

- While conventional farming entails an intensification of agriculture by chemical means, organic farming intensifies farming by biological and ecological means generated mainly on the farm itself. Sustainable land use aims to prevent the overexploitation of soil, organisms and resources.
- A high level of food quality is the basis of healthful nutrition. High quality means tasty products, high levels of beneficial substances, low levels of harmful substances and, at first, when refrigerators were not yet common, food that keeps well for a long time.
- Organic farming was always connected to a vision of 'another', 'better' society: during the pioneer period, self-sufficient gardeners leading a 'natural' lifestyle; later, the preservation of a vanishing rural culture; today, a vision of a sustainable society.

Nevertheless, there have been several important changes since the beginning of organic farming in the 1920s:

- Some core ideas of the pioneer period were abandoned, such as vegetarianism, going back to the land and an economy based on recycling municipal waste and humus.
- On the other hand, organic farming incorporated new core ideas: appropriate animal husbandry, environmental protection and the renunciation of genetically modified organisms.
- Many innovations improved organic farming practice in aspects of cropping such as composting, soil cultivation, rotations and weed management, as well as appropriate animal husbandry (see Chapter 5, this volume).

Some issues have remained unresolved:

- From the beginning, organic farming called for a variety of regional cultivars fitting organic farming conditions; serious efforts to breed such plants began only recently.
- The pioneers of organic farming tried to compensate for the loss of minerals from harvest, erosion and leaching by recycling urban wastes and building up a municipal waste and humus economy. Continuously increasing yields must have aggravated the problems caused by the lack of such recycling. However, use of sewage sludge, a major potential source of nutrients, is not allowed in organic farming, although some European farmers are getting involved in municipal waste composting schemes.
- Although food quality is a core topic of agriculture, there exists insufficient knowledge on how to improve food quality apart from not using mineral fertilizers and synthetic pesticides.
- If agricultural ecosystems are viewed as 'cultivated' nature – as opposed to 'natural' nature – the question of how to reproduce and

improve such 'artificial' states of nature must be answered; corresponding agroecological concepts which emphasize the reproduction of the whole farm to prevent overexploitation do not yet exist, except in special cases such as permaculture.

Having undergone the historical developments traced in earlier chapters, where are we heading now? The organic world has matured, getting involved in every part of the food sector: politics, national and international trade, research, public relations and the consumer world. Organic farming is no longer controversial, organic labels have become much more familiar and many organic farming organizations are recognized as capable lobbyists and as good sources of knowledge and analysis.

But these extremely positive results of several decades of hard work have also made the world of organics very complicated and difficult. It takes real specialists and experts to penetrate the latest European Union legislation, standards and rural programme proposals. Luckily, the organic movement has fostered such experts, who today are well regarded. But it is difficult to keep the grass roots movement – the farmers, consumers, advisers and researchers – updated and involved. A crucial task is to keep some initiative within the movement itself, to be able to develop organic agriculture in accordance with the principles and values formulated and decided by the whole organic world together.

An important role in keeping organic farming dynamic will be played by research, and a major challenge for future organic farming research will be to breathe life into the well-worn concept of sustainable food and farming systems. An increasing number of research projects and meta-studies have recently been dedicated to many aspects of organic farms, including soil conservation and soil fertility, environmental protection, biodiversity, wildlife and landscape (El-Hage Scialabba and Hattam, 2002; Mäder *et al.*, 2002; Hole *et al.*, 2005; Pimentel *et al.*, 2005). Organic food systems are more successful than others in combining productivity and sustainability (Mäder *et al.*, 2002; Pimentel *et al.*, 2005). In the context of rural development, organic farming offers an attractive chance for many regions around the world. Organic farming, especially in the tropics and subtropics, can improve the performance and the stability of marginal and subsistence food production in a sustainable way. A limited amount of scientific data – not much has been generated so far – shows organic farming's power to overcome starvation and poverty (e.g. Pretty *et al.*, 2003). A major focus of future research will be organic farming's potential to contribute to policy makers' worldwide strategic objectives for agriculture and food systems. This will also shift the priorities of the whole organic community away from the technical aspects of quality assurance that have ruled the development of organic farming during the last 15–20 years, a focus dominated very much by the needs of consumers in the north.

Organic farmers must have an important part in such research. Organic farmers were once active and independent, although sometimes awkward, pioneers. The long process of standardization of legal and private requirements and procedures tended to reduce them to passive objects. This process of harmonization has been very important to gain the trust of consumers and policy makers. To cope with the challenges described above, inspected and certified organic farmers will again have to become entrepreneurs and visionaries. Researchers will have to respect this and integrate farmers as subjects into their scientific work. In natural resource management, nature conservation and rural development, it is now recognized that research and development can no longer be the exclusive work of scientists (Gottret and White, 2001; Gonsalves et al., 2005). This is where organic farming research should strive for mastery (Alrøe and Kristensen, 2002; Baars, 2002).

Eventual fulfilment of the potential of this promising form of partnership will in large part fall to the new generation of organic farmers that is slowly taking over. This succession is a very interesting and positive change. The pioneer organic farmers and researchers are approaching retirement or have already retired. In the best cases the pioneers' children take over their farms, bringing new visions for organic production, creative suggestions about research, new ideas on how to make farms survive economically and new thoughts on creating a viable, high-quality lifestyle. It is an exciting time indeed! Equally exciting are the new market trends, where a new interest in global issues is taking shape, strongly anchored in the young generation. New concepts are promoting the quality of organic food among young consumers. The new trend is about the good life, where health, taste, exclusiveness, origin, solidarity and fairness are key concepts. If the organic movement learns how to navigate these trends, it has a great chance to win a new generation of consumers and at the same time deepen and develop the organic concept.

If the organic concept indeed is to be deepened and developed, then among the distinct but related issues that must be dealt with are those of globally versus locally oriented organic production, ethical and social justice considerations in organic trade and the role of large-scale corporations in the organic world of the future. We also need to critically discuss the question of whether expanding the market for certified organic products should be an aim in itself, or rather a means to other ends, e.g. environmental and health benefits or greater food security in developing countries; in the latter case, this can be achieved by farms that use organic methods but are not certified.

The rapid growth and worldwide trade in organic products is a reality. This is not just significant for commercial enterprises, but may also offer small farmers in developing countries the opportunity to sell their products at an appropriate price. But the worldwide movement of goods also presents a challenge for the principles of organic farming: can we arrange the expansion and globalization of the organic sector without compromising the values that have characterized organic farming as an alternative economic approach?

With worldwide sales of organic products approaching US$30 billion, organic farming is no longer just a niche business. It is not just products such as coffee and tropical fruits that find their way to our tables from all corners of the earth, but also organic soybeans, vegetables, wines and many others. The processing and trade of organic foods is now handled by a large-scale and very much mainstream industry. Whether or not one welcomes this development, it is noteworthy that even McDonalds includes organic foods in its marketing strategy, having offered organic milk in Sweden and the UK for many years. Some early organic companies have expanded greatly, with annual sales now reaching hundreds of millions of dollars. Others have been acquired by multinational corporations, while keeping their original brand names. It is likely that many organic consumers typically do not realize that their purchases are contributing to the profits of corporations such as Nestlé, Coca-Cola, Unilever, Kraft and Cargill.

Parallel to the development of multinational organic food producers, similar trends are evident in food retailing. Supermarket chains of all sizes now sell organic products, which account for as much as 10% of the sales of such giants as tegut in Germany or Coop in Switzerland.

It is not surprising that the organic sector has undergone this development. Organic food companies also work within a capitalist context. Given the sustained, profitable growth of organics, who can seriously expect that large food companies would ignore the opportunities? In its emergence from niche status, is it at all likely that the sector would escape the prevailing economic system (even if it should wish to do so)?

There are those who welcome this development and hail it as a success story. Others will accept it fatalistically. But the credibility of organic farming is at stake, which calls for critical reflection and debate, as well as intensified efforts to develop alternative markets, such as farm stands, farmers markets, whole food shops and community-supported agriculture.

The best answer may lie in maintaining a range of marketing options so that organic farmers on one hand are not overly dependent on multinational supermarket chains, but on the other are not restricted to a fashionable but small niche that would allow only a small minority of producers to be organic.

Collaboration with Fair Trade organizations is critical here. The recently founded venture 'Bio, regional and fair' is an example of an effort to counter globalization. This Bavarian organization (described at www.bio-regional-fair.de) brings together fair trade groups, consumer associations, church organizations, regional initiatives and organic farmers, all of whom realized that they were pursuing the same basic goals: to enable farmers to earn a fair income that secures their livelihood; to strengthen regional economies; and to protect nature and the environment.

Although 'trade' is a bad word in some organic circles, from the viewpoint of developing countries' farmers there are strong arguments in favour of international trade in organic products. (So, too, from the viewpoint of the consumer. Until some decades ago, bananas, coffee and chocolate were still relative luxuries. Yet today, more coffee is consumed in Germany than beer, which shows how unrealistic it is to dogmatically demand that one should only buy local and regional products. Why should we not be allowed to appreciate an organically grown banana or a cup of fair trade coffee or a bar of chocolate?)

For many developing countries, exporting agricultural products is the only way to participate in international trade. Thanks to their geographic advantages and their decentralized, small-scale farming systems and low labour costs, these countries can produce food and agricultural raw materials at competitive prices. Moreover, because of their tropical and subtropical climates they can produce many products that do not grow in the temperate zones. The higher prices earned for organic products are particularly significant for farmers in these countries; especially when combined with the higher prices received with fair trade, this frequently means survival in the true sense of the word.

In many developing countries, conversion to organic farming occurs not just for export crops such as coffee or tea, but for all crops over entire regions. This means that farm families and consumers in the area can enjoy high quality, organically grown food. However, the greater purchasing power of consumers in rich countries often makes it difficult, if not impossible, to sell significant quantities of organic products in developing countries, where the organic price premium can be enormous. For example, in China organic vegetables can cost five times as much as conventional ones, and this in a country that is having a big impact on world markets for organic food because of its low production costs and wages. Yet at the same time, a great potential role for organic farming may not involve markets at all; rather it is to offer highly productive subsistence farming systems in developing countries, protecting the health and environment and enhancing the food security of millions of poor farmers around the world.

Consumers who buy organic food typically do so for health reasons, and may not care whether it was transported a long distance or whether it was produced under socially acceptable conditions. But an increasing number want to know where their food comes from and the conditions under which it was produced. Fortunately, therefore, there is an increasing demand for products certified as both organic and fairly traded.

Organic regulations do not limit the distance the food travelled, its seasonality or the energy used in processing and transporting it. However, these issues feature in public debates, and we are beginning to see regulations applying to the social aspects of agriculture. But we need to be wary of letting the notion that 'local is best' degenerate into a mindless, absolutist slogan that trumps all other considerations; rather, we must critically examine when, how and why it is best, and when, how and why it is not. Moreover, if support for locally produced food is to be a satisfactory basis for designing food systems for the future, it needs to be complemented by organic, fair trade and ethical principles.

It is becoming increasingly urgent to consider the 'ecological footprint' of food products, i.e. the resources and land required throughout the entire production process. With its excellent systems of certification, organic agriculture can provide complete traceability of food from the package back to the origin. The 'Nature and More' system (available at: www.natureandmore.com) has created the basis for giving consumers maximum transparency concerning the movement of goods and their production conditions, so that anyone who wishes can obtain excellent information and make their consumption decisions in accordance with their values.

The international organic movement, in particular the International Federation of Organic Agriculture Movements (IFOAM), has not only been addressing the phenomenon of globalization in the abstract, but has also been developing numerous projects to boost local and regional marketing in developing countries. Amazing results have been achieved with dedicated, committed effort. For example, the biodynamic Sekem project in Egypt has succeeded in making its range of organic teas the market leader in this nation of tea drinkers. Sekem initially exported 80% of its organic products, but now sells that share domestically. Its remarkable success earned Sekem the 'Alternative Nobel Prize' in 2003, particularly in recognition of its pioneering achievements and innovative strategies in marketing organic products, which are making an important contribution to the local economy. The vision and principles underlying Sekem prompted IFOAM to develop a code of conduct and gradually implement it.

The slogan 'think globally, act locally' is popular in environmental circles, but as is often the case with slogans, it falls short. Should we really leave global action to the World Trade Organization and multinational corporations? And how successful can eating locally be if we do not think about it? Inevitably, we must think *and* act locally, regionally and globally.

In thinking about the future of organic farming at whatever geographic scale, a broader and more comprehensive description of the current organic market is needed to establish a more holistic baseline so that future changes can be evaluated. Measures of growth, such as estimated sales, cultivated area and growth rates, are relatively straightforward and available. However, additional measures are needed to assess changes in the character of organic agriculture, such as the evolution of organic standards and certification processes, the status of price premiums and how the proceeds from organic food sales are divided among farmers, farm workers, processors and retailers. The evolving terms of farmer/buyer contracts, what kinds of farmers have access to markets and why, and the connections between organic agriculture and local food production, fair trade and farm workers' rights are all values-based choices that the global organic community is now facing. Additional criteria are needed to evaluate more fully which factors contribute positive benefits and who wins and who is hurt by these structural and institutional changes.

How these changes affect the social, as well as the environmental and economic sustainability of organic farming is also not fully understood. Many writers are concerned about changes in the structure of the industry as giant food corporations acquire organic companies or develop their own organic divisions, as described earlier. Others worry about increased vertical integration and the possible 'commodification' of organic products. A related concern that is frequently voiced – correctly or incorrectly – is that new organic farmers have come in just to take advantage of the economic opportunities created by the growth of the global organic market, and are therefore less committed to organic principles and values. But rather than simply bemoaning this development and dismissing the newcomers as 'crypto-conventional', it would be much more constructive to educate them on organic principles so that they become more committed supporters.

The fact that in 1996, IFOAM drafted a chapter on social justice in its production and processing standards that broadly recognizes the needs and rights of farm workers and all other people involved in organic production and processing is a very important step towards addressing these looming challenges. In 2003, the Soil Association (no date) in the UK launched optional 'ethical trade' standards to address the growing interest in including fair trade in organic agriculture in the north, and Rural Advancement Foundation International, USA, issued a call for social stewardship standards that has helped launch the North American development of social justice claims for organic farmers (Henderson *et al.*, 2003). Many people clearly understand that if organic farming follows conventional agricultural relationships, especially those based on economic exploitation, a major part of the collective set of organic values will have been lost. However, it is very difficult to draw clear lines in such discussions. Just because a farm or company is big or is owned by a multinational corporation, does that automatically mean it cannot be truly organic?

Working to maintain the integrity of organic farming during its recent phase of institutionalization is very difficult, because it is about an approach to agriculture that was initially developed and promoted mainly by private, grass roots groups, but which now faces formal national and even international regulatory oversight. Governments and large multinational corporations in both hemispheres are getting involved. There is concern that because of growing policy interventions, at least in Europe, the organic sector has lost control over its own destiny and that policy makers are now writing the rules, perhaps trying to accommodate the needs of large corporations and free trade rather than the principles put forth by the pioneers of the organic movement. Bilateral and multilateral trade arrangements, national and state laws, Codex Alimentarius organic labelling standards, as well as corporate concentration, mergers and buyouts are having daily impacts on the organic value system. All public and private stakeholders, including farmers, labourers, processors, handlers, retailers, suppliers and consumers, must actively participate in the debates about these changes. They must recognize that governments cannot and will not solve all the problems that come with growth, success and the entrance of new players. With organic farming, unlike many publicly funded schemes (especially agri-environment programmes), the 'rules' were originally developed largely through a grass roots process. Neither do state officials, food companies nor researchers own these rules, which makes some of them uncomfortable with the organic concept. The state, corporations and the research community must learn to live with this, and develop new partnership models involving all interested parties, rather than seeking to dominate the process.

Will this new phase require a new movement to pick up the pieces because organic farming has failed to preserve its integrity? Or will this next phase help a broad and generous vision of organic farming to blossom, with environmental stewardship, improved quality of life and social justice as realities? These are fundamental values that are still worth fighting for: an ongoing struggle for food with a place, a face, a taste and an attitude. History will judge organic agriculture both by what it becomes and by what values were allowed to fade away.

References

Alrøe, H.F. and Kristensen, E.S. (2002) Towards a systemic research methodology in agriculture: rethinking the role of values in science. *Agriculture and Human Values* 19, 3–23.

Baars, T. (2002) Reconciling scientific approaches for organic farming research. Part I: Reflection on research methods in organic grassland and animal production at the Louis Bolk

Institute, The Netherlands. PhD thesis. Publication Nr. G38, Louis Bolk Instituut, Driebergen, Utrecht, The Netherlands.

El-Hage Scialabba, N. and Hattam, C. (eds) (2002) Organic agriculture, environment and food security. Chapter 2. *Environment and Natural Resources Series* 4. Food and Agriculture Organisation of the United Nations, Rome.

Gonsalves, J., Becker, Th., Braun, A., Campilan, D., de Chavez, H., Fajber, E., Kapiriri, M., Rivaca-Caminade, J. and Vernooy, R. (2005) *Participatory Research and Development for Sustainable Agriculture and Natural Resource Management: A Sourcebook. Volume 1: Understanding Participatory Research and Development.* International Potato Centre – Users' Perspectives with Agricultural Research and Development, Laguna, Philippines, and International Development Research Centre, Ottawa, Canada.

Gottret, M.A.V.N. and White, D. (2001) Assessing the impact of integrated natural resource management: challenges and experiences. *Conservation Ecology* 5, 17. Available at: http://www.consecol.org/vol5/iss2/art17/

Henderson, E., Mandelbaum, R., Mendieta, O. and Sligh, M. (2003) Toward Social Justice and Economic Equity in the Food System, A Call for Social Stewardship Standards in Sustainable and Organic Agriculture. Final Draft, October. Rural Advancement Foundation International – USA. Pittsboro, North Carolina. Available at: www.rafiusa.org/pubs/SocialJustice_final.pdf

Hole, D.G., Perkins, A.J., Wilson, J.D., Alexander, I.H., Grice, P.V. and Evans, A.D. (2005) Does organic farming benefit biodiversity? *Biological Conservation* 122, 113–130.

Mäder, P., Fliessbach, A., Dubois, D., Gunst, L., Fried, P. and Niggli, U. (2002) Soil fertility and biodiversity in organic farming. *Science* 296, 1694–1697.

Pimentel, D., Hepperly, P., Hanson, J., Douds, D. and Seidel, R. (2005) Environmental, energetic, and economic comparisons of organic and conventional farming systems. *BioScience* 55, 573–582.

Pretty, J., Morison, J.I.L. and Hine, R.E. (2003) Reducing food poverty by increasing agricultural sustainability in developing countries. *Agriculture, Ecosystems and Environment* 95, 217–234.

Soil Association (no date) Soil Association ethical trade. Available at: www.soilassociation.org/web/sacert/sacertweb.nsf/B3/ethical_trade.html

This page intentionally left blank

Index

Accreditation
    Australia 235, 236
    historical development 159
    IFOAM 170, 171, 181–182, 232, 259
    international 168–169
    ISO Guide 61, 257
    private 169–170
    and reconciliation of standards 255
    US Department of Agriculture 156–157, 261
    see also Certification
Agricultural bacteriology 11–12
Agricultural research
    Albert Howard 24, 45–46
    biodynamic 42–45, 51
    institutions
        Danish Research Centre for Organic Food and Farming (DARCOF) 54
        Louis Bolk Institute 52
        Ludwig Boltzmann Institute for Organic Agriculture and Applied Ecology 53
        Organic Research Centre (Elm Farm) 53
        Research Institute of Organic Agriculture (FiBL) 51
        Rodale Institute 51
        US Land Grant Universities 54–55
    organic *see* Organic farming, research and development
    Robert McCarrison's on nutrition 25
    specialized and reductionist versus holistic 41–42, 66–67
    whole farm 24, 41
Agriculture biologique 17, 96
American Association for the Advancement of Science (AAAS) 1
Animal diseases
    allowed allopathic medicines against in organic farming 83, 84
    on organic and conventional farms 84
    preventive strategies 64–65, 83–84, 250
Animal production
    absence in Natural agriculture 14–15, 16
    animal welfare 65, 75, 250
    attempts to reduce antibiotic use 65
    breeding for organic 87
    complementary medicine 250
    conflict with organic principles 85
    control of parasites in organic 84–85
    diversity of 64
    feeding of roughages in organic 65–66
    in Australia 226
    in Haughley experiment 48, 49
    homoeopathic treatment 84
    late start of organic research and standards 63
    organic concepts of health 85–86

275

Animal production (*Continued*)
  public reaction against industrialized methods 82–83, 161
  role in organic farming 45, 75, 159, 161
  subsidies and support payments 104, 105
Anthroposophy *see* Biodynamic agriculture
Argentine Movement for Organic Production *see* MAPO
Aubert, Claude 17, 155, 178

Balfour, Eve
  Haughley experiment 25, 42, 47–50, 75, 189
  influence on organic farming 31, 41–42, 51, 52, 59, 66, 177, 192
  Soil Association 25, 135, 187–190, 193–196
  The Living Soil 35, 47, 188
Bebauet die Erde 15–16
Bergland, Bob 2–3
Biodynamic agriculture
  animals' place in 75, 82
  and anthroposophy 19–20
  Australia 228, 231, 236
  concepts 6, 19–20, 162
  Demeter
    journal 22
    label 22, 117, 123, 153
  food quality 21
  in Third Reich 21, 22–23
  origins 19–20, 42–43, 176
  picture-forming methods 41, 248–249
  plant breeding 20, 24, 86–87
  post-World War II development 23–24
  preparations 19, 20–21, 23, 44, 78
  research
    animal health 44
    approaches 41
    centres 51
    DOK trial (FiBL) 52, 55–56, 247–248
    preparations 21
    nutrient uptake 60–61
  Sekem project (Egypt) 270
  see also Steiner, Rudolf
Bourgeois, Denis 177–178

Bromfield, Louis 26
Butz, Earl 2–3

Carson, Rachel 4, 34, 190
Certification 96–97
  and agri-environmental support 101, 116
  Argentina 219, 222
  Australia 227, 230–233, 234–236, 237, 239
  consumers' knowledge 138
  diversity and inconsistencies amongst programmes 182–183, 261
  European-wide system 169
  EU support 113
  eventual need for 152
  historical development 159–160
  importance for building trust 141–143
  in developing countries 168–169
  KRAV 212–213
  mutual acceptance among certifiers 170
  Organic Foods Production Association of North America 253–255, 256–257
  overburdening of certifiers and farmers 170, 171
  social justice standards 36
  Soil Association 197–199
  Switzerland 243, 251
  and traceability of food 270
  USDA-accredited 156–157
  see also Accreditation; Organic standards
Chavez, Cesar 34
Chevriot, Roland 177–178
Codex Alimentarius
  coordination with other organizations 155, 171, 236, 251, 259
  Guidelines on Organic Agriculture 35, 83, 131, 157–158, 170, 182
  principles of organic farming 30–31, 168

Danish Research Centre for Organic Food and Farming (DARCOF) 54

Demeter *see* Biodynamic agriculture
Diseases
    Albert Howard's writings on 46, 74
    occurrence in nature 74
    see also Animal diseases; Human diseases; Plant diseases

Ecological Farmers Association (Sweden)
    activities 206–209
    and the Ecological Forum 211–212
    goals 206
    origin 204–205
    policy work 208–209, 212
    relationship with KRAV 212–213
    relationship with Swedish Farmers Association 213–214
Ekologiska Lantbrukarna *see* Ecological Farmers Association
European Action Plan 98, 112–113, 117, 133, 273
European Union (EU)
    Common Agricultural Policy (CAP) 202
        agri-environmental schemes 100–102, 103–105, 109, 111, 115
        effects on organic farming 94, 111, 117, 194
        information-related support 107–110
        price and income supports 103
        processing and marketing support 105–107
        reforms 100, 103–104, 105, 117, 132, 198, 202
        rural development programme 110–111, 207
        support for organic farming 99, 132
    LEADER programme 110–111
    logo for organic products 156
    networks on organic animal production 64
    organic imports into 234
    recognition of equivalent countries 219
    regulations on organic farming 159, 183–184
        animals 65
        crops 164
        first draft 183–184
        revision 156, 167
        requirements on certification bodies 169
    research funding 54
    Structural Funds 111
    support programmes 209, 210, 211

Farming Systems Trial (Rodale Institute) 51, 56–57
Faulkner, Edward H. 26
Fertilizers
    in biodynamic agriculture 44, 137, 153
    Chilean nitrate 164–165, 257
    effects on food quality and human health 11, 44, 58, 61, 189, 247
    effects on plants and soil 10, 11, 25, 60, 188
    in Haughley experiment 48, 49, 189
    in natural agriculture 15
    municipal wastes 14
    Nazi view of 22
    non-use of mineral/synthetic in organic farming 30, 55, 73, 76, 154, 176
    organic compared with mineral 12, 44, 50, 58, 60, 61
    rock powder 17, 243
    water pollution and leaching 4, 57–58, 202
FiBL (Forschungsinstitut für biologischen Landbau) 51–52, 242
    animal welfare 250
    biodynamic research 248–249
    early research 247–249
    founding 245–246
    information activities 107
    relationship with IFOAM 251
    standards and logo 251
Food Reform 12–13, 26
Fungi
    and nutrient cycling in soil 47
    mycorrhizal 11, 46, 47, 56, 77
    pathogenic 74, 80–81, 87
    plant defences 80–81, 248
Fungicides
    allowed in organic farming 75, 243
    copper 81, 164, 257

Fungicides (*Continued*)
   non-use in organic seed production 80
   see also Pesticides

Genetically Modified Organisms (GMOs)
   contamination by 164, 221
   organic agriculture as the alternative 37
   prohibited in organic farming 132, 133, 159, 163, 164, 197, 238, 257
   suit against in Argentina 220–221
Gesamthochschule Kassel 53
Goetheanum (Dornach, Switzerland) 51, 243
Görbing, Johannes 12, 17

Hensel, Julius 17, 243
Herbicides
   alternatives in organic farming 79
   and tillage 77, 230
   labour saving 78
   nitrofen contamination scandal 142
   see also Pesticides
Hofstetter, Mina 17, 243
Howard, Albert
   ideas
      on agribusiness 33
      on agricultural research 24, 41–42, 46
      on agricultural systems 46, 74, 264
      on farmers' knowledge 32, 45
      on health of soils and plants 24, 46, 47, 59, 73–74, 79, 188–189
      on soil organic matter and composting 45–46
   influence on organic farming 25, 31, 51, 66, 82
   and Soil Association 189, 190
   work in India 24, 45–46, 188–189
Howard, Gabrielle 24
Howard, Louise 24
Human diseases
   and diet 13, 46
   and health 25
   and soil 74
Hunzas (India) 25, 47, 189

Insecticides
   natural 75, 81
   origins in World War II 5
   see also Pesticides
Insects
   attacks on unfit plants 10, 78
   crop resistance 59–60
   diversity on organic and conventional farms 58–59
   natural controls 58–59
Integrated organic action plans 111–113
   see also European Action Plan
International Organic Accreditation Service (IOAS) 182, 236
International Federation of Organic Agriculture Movements (IFOAM)
   accreditation of certifiers 169–170, 232
   aims and activities 175–176
   basic standards 38, 97, 131, 154–155, 158, 161, 164–167, 182
   cooperation with public and private bodies 97, 156, 170, 183–185
   eligibility of members from industry 179–180
   founding and early years 35, 176–178, 251
   growth and diversification 3, 178, 180
   organic guarantee system 181–183
   organizing in North America 254, 259
   principles of organic agriculture 36, 166–167, 170–171
   publications 180
   scientific conferences 2, 52–53, 63, 180, 214, 219, 251
   smallholder certification 183
   social justice standards 271
   support for marketing projects 270
   study of energy use and climate change 37

Keene, Paul 32
King, Franklin H. 13, 31–32,
Könemann, Ewald 16–17, 243
KRAV (Sweden) 205–206, 207, 208, 209, 211, 212–213, 214

Landreform 26, 236–237
Langman, Mary 177, 196

Lebensreform 12–14, 16–17, 21, 26, 243
Liebig, Justus von 40
Life Reform *see Lebensreform*
Löhnis, Felix 11–12
Louis Bolk Institute 52
Ludwig Boltzmann Institute 53
Lymington, Viscount 189

Manure
    and soil fertility 12
    animals as providers 58, 61, 75, 250
    composting 15, 248
    effect on food quality 60–61
    efficient use of 76
    in biodynamic agriculture 44, 153
    in natural agriculture 15
    in the Haughley experiment 48, 49
    nitrogen leaching 57–58
    quality 44
    restrictions on use 256
MAPO (Movimiento Argentino para la Producción Orgánica)
    goals and activities 219–220, 223
    members 219
    suit about GMOs 220–221
McCarrison, Robert 11, 25, 47, 61, 189, 192
Miliband, David 200
Milton, Reginald 50, 191
Mother Earth 25, 35, 187, 190, 191–192, 194, 198
    see also Soil Association
Müller, Hans
    as founder of organic–biological agriculture 18, 152–153, 154
    as member of Swiss Parliament 243
    conflict with Swiss research establishment 243–244, 246
    lack of interest in animal welfare 250
Müller, Maria 18, 242, 243–244, 246

National Association for Sustainable Agriculture, Australia (NASAA)
    accreditation 232, 236
    aims 228
    certification 235–236
    education 231
    marketing 231–232
    organic standards 229–230, 234
    origin 227–228
    policy influence 232–233
    unifying efforts 236–237, 239–240
Natural agriculture 14–16, 243
Nature et Progrès (France) 17, 154, 177
Network for Animal Health and Welfare in Organic Agriculture (NAHWOA) 64
National Organic Program (US Department of Agriculture) 131–132, 156–157
National Organic Standards Board (US Department of Agriculture) 256–257

Organic agriculture *see* Organic farming
Organic-biological agriculture 18–19, 152–153, 154, 203
Organic farmers
    and animal welfare 82
    as innovators and researchers 12, 18, 31, 63, 83, 251, 267
    consumers' trust in 142–143, 152
    declining number in Argentina 220
    developing countries 182–183, 269
    'hippies' 5, 33, 209
    increasing number in Australia 226
    informal and experiential knowledge 41, 67, 76, 84–85, 87
    new entrants 6–7, 115–116, 194–196, 267
    reasons for entering or leaving organic 5–6, 33, 78, 95, 102, 104, 202, 271
    relationships with conventional farmers 213–214
Organic farming
    and alternative lifestyles 5
    Argentina 221–223
    as alternative to industrialized farming 30, 32–33, 36, 73
    Australia 226–227
    beginnings in India 24–25
    changing image 1–2, 6, 34
    early organizations 176–178
    education 231
    effects of globalization 35–36, 268
    emergence in UK and USA 25–27

Organic farming (*Continued*)
  and environmental movement 4–5, 218
  and natural agriculture 14–15
  and organic-biological agriculture 18–19
  evironmental stewardship 37
  Far East influence 13 14, 31–32
  fertility-building crops 77
  global spread 3
  growth 93–94, 113, 125
  indigenous knowledge 31–32, 217
  information-related public support
    EU-level 108–110
    national 107–108
  Life Reform and Food Reform as precursors 12–13
  'mainstreaming' 66, 252, 266, 268
  policy support
    benefits and drawbacks 113–115, 117
    effect of Common Agricultural Policy 103–105
    financial 98–102, 211
    information-related 107–110
    integrated action plans 111–112
    justification 94–96
  principles and values 31–36, 165–168, 264–265
  reasons for emergence 9–11
  research and development
    animals 63–66, 75, 82–85, 250
    Australia 238–239
    biodynamic 42–45, 50, 87, 247
    early 11–12, 42–50, 190–191, 243–244
    consumers and markets 137–138, 140, 250
    crop production 76–82
    financial support 2, 108–109
    food quality 60–61
    institutions 50–55
    major themes 55–56
    natural pest control 58–59
    nutrient losses 57–58
    participatory 41, 251, 267
    rapid growth 2
    relationship to mainstream science 40–42, 46–48, 66–67, 88, 243–246
    resistance to pests 59–60
    soil ecology 55–57
    Sweden 207
    Switzerland 243–246
    weeds 62–63
  resistance and criticism 2, 214–215, 244–245
  rural development 110–111, 266
  social justice 37–38, 271
  Sweden 203–204, 209
  targets 210, 211–212
  see also Animal production
Organic Federation of Australia 237–239
Organic foods
  and fair trade 37–38, 127–128, 130
  and health 134
  and locally grown foods 143–144, 146, 269, 270
  availability 134–135, 145
  characteristics of consumers 136–137
  consolidation of markets 134
  consumers' interest and attitudes 96, 136, 140–141, 144, 147
  consumers' knowledge 137–138, 146
  demand for 135–136
  differentiation amongst 130
  direct sales 146
  distribution amongst countries 125–126
  early markets 18, 32, 33, 123, 127, 134
  frequency of consumption 138–139
  in Life Reform 13
  involvement of large corporations 33, 134, 268
  legislated definitions 117, 234–235, 256–257
  logos and labels 98, 123, 136, 141–143, 156, 238
  major categories 139
  market development 231–232
  'myth' 1
  prices 144–145
  processed 47, 130, 142, 147, 258–259
  public procurement 107
  quality 60–61
  sales by major retailers 124, 128–129, 146, 197–198, 268
  specialized shops 127–129
  supply–demand imbalances 105, 106–107, 114–115
  trade 1, 35, 124, 267–269

see also Organic standards
Organic Foods Production Act of 1990 (OFPA) 257, 260–261, 262
Organic Foods Production Association of North America (OFPANA) see Organic Trade Association
Organic Research Centre (Elm Farm) 53
Organic standards
  advantages and disadvantages of legal establishment 97–98, 117, 131–132, 133, 273
  animal production see Animal production
  Codex Alimentarius Guidelines see Codex Alimentarius
  concern over weakening and loss of integrity 6–7, 35, 199
  consumers' knowledge and attitudes 138, 143, 146
  discrepancies and need for harmonization 35
  earliest 15, 22, 153–154
  European Union 97–98, 117, 131, 156
  historical development 159–160
  IFOAM Basic Standards see International Federation of Organic Agriculture Movements
  Japanese Agricultural Standards 131–132
  lack of early need 152
  legislation establishing 96–98, 155
  social justice 38
  Soil Association 154, 194
  US Department of Agriculture see National Organic Program (USDA)
  see also Certification; Organic foods
Organic Trade Association (OTA)
  approved materials 256–257, 258–259
  Ethical Practices Committee 259
  internal tensions 260–261
  international linkages 259
  origin 254
  promoting organic farming 259–260
  reciprocity amongst certifiers 260
  response to draft federal standards 262
  standards 254–25, 258–259

Pesticides
  acceptance in Argentina 217
  allowed in organic farming 75, 243
  and food safety 172, 202
  and intensification of farming between world wars 10
  banning of organochlorines 4, 75
  campaign against 196, 197
  consumers' concerns and knowledge 34, 86, 138, 140, 172, 197, 247
  environmental impacts 34, 86, 202
  health hazard to workers 34
  incorporation into plant via genetic engineering 86
  non-use in organic production 30, 73, 154, 176
  'residue-free' foods 257
  residues 11, 61, 161
  spill into Rhine 249
  see also Fungicides, Herbicides, Insecticides
Pests
  biocontrol agents 81, 249
  control in nature 74
  habitat management and redesign of systems for control 79–80, 251
  inappropriateness of chemical control 80
  Integrated Pest Management 245
  likely increase 80
  modern control techniques 80–81
  plant defences 81–82
  see also Fungi; Insects; Plant diseases; Weeds
Pfeiffer, Ehrenfried 22, 41, 43–45, 243
Plant diseases
  and marketability of crops 81
  and soil fertility 81, 189
  as sign of inappropriate management 79
  chemical control of in organic farming 75
  modern organic control methods 80–81
  on organic and conventionally grown crops 60
  resistance 59, 81–82

Research see Agricultural research
Research Institute of Organic Agriculture see FiBL

Rodale Institute 51, 56
Rodale, Jerome I. 26, 31, 32–33, 51
Rothamsted (UK) 42, 46, 48, 49
Rule [Law] of Return 46–47, 189, 192
Rusch, Hans Peter 18, 203, 242, 244–245, 246

Schalcher, Heinrich 246
Schumacher, E.F. 190, 193, 196, 197
Soil Association (UK)
    and founding of IFOAM 176–177
    and growth of UK organic market 197, 198
    early organic standards 154, 194
    E.F. Schumacher as president 193
    Eve Balfour's role in founding and growth 188, 189, 192
    founding 34, 136, 187–190
    generational clash in 1970s 194–196
    Haughley experiment 49, 190–191
    market orientation 194–196
    Mother Earth 191–192
    promotional and educational activities 192–193
Soil erosion 1, 2, 15, 26, 45, 222
Steiner, Rudolf
    Agricultural Lectures 19, 43, 176
    ideas about agriculture 19, 20, 41, 42, 43
    influence on organic farming 20, 23, 31, 44, 153, 203, 242
    see also Biodynamic agriculture
Sustainable Agriculture Farming Systems project (University of California) 54
Sustaining Animal Health and Food Safety in Organic Farming (SAFO) 64

Tillage 14–15, 74, 77–78, 230, 250

United Nations Conference on Trade and Development (UNCTAD) 3, 160, 170, 171, 183, 184
United States Department of Agriculture (USDA) 2, 156–157, 258, 259, 261–262
    see also National Organic Program; National Organic Standards Board
University of Kassel 53

Vogtmann, Hartmut 53, 154, 245, 251

Waksman, Selman A. 12, 26–27
Weeds
    ecology 62–63
    novel control techniques 78–79
    problem in early organic farming 74
    reduced requirements for control 78
    species diversity 62–63
    tillage for control in organic farming 77–78
Wheel of health 25
Williamson, George Scott 189, 190
World Trade Organization (WTO) 252, 270
    EU and Codex standards as references 131, 170
    harmonization and equivalence of regulations 155, 157, 170
    impact on Australian organic regulations 235
    tensions with EU 105